西餐料理 一把罩

精進版

作者序

　　中國人的對飲食的傳統觀念裡向來是"民以食為天"，又說到開門七件事：柴、米、油、鹽、醬、醋、茶，都跟吃有關。而且西方世界對於"吃"的飲食文化，與華人相較之下，亦不惶多讓，更加說明了人類的進化，與延續生命的過程中都離不開"吃"，而且這是一個放諸四海皆準的道理。

　　中西方飲食文化最大的差異即在於，烹調的技巧、食材的選用、辛香料調味及廚具器皿使用等。在西方飲食往往給予大家製作煩瑣，食材特殊不易取得等因素，因此在國人的飲食與烹調習慣中較不易受到青睞。但是在凡事講求效率的資訊化時代，一切講求新、速、實、簡，因此西方餐飲文化已經漸漸的融入國人的飲食習慣之中，進而西餐料理也成為國人的日常。

　　或許大家會認為西餐給人華而不實的錯覺，為了推翻上述偏差的論調，筆者針對國人的飲食習慣，加上本身對於廚藝的樂衷與修習西式廚藝的教學經驗，將自身所學融會貫通。希望將西式餐飲帶入國人的生活之中，讓大家瞭解西餐其實是高貴而不貴，兼具美味及營養，並且可以全家DIY。西餐料理一把罩精進版，內容更充實，作法更Easy，「防疫新生活，在家作料理」，在此與廣大的讀者分享。

推薦序

　　周景堯教授（Andy Chou）是國內少有精通中西菜系、又有超人管理能力的「廚藝管理」和「廚藝教育」專家。他早年留學加拿大，學習西方廚藝；學成歸國又在五星級酒店由基層做到主廚，業界磨練多年後轉換跑道，在經國管理暨健康學院擔任餐旅廚藝管理系主任。任職十餘年，為臺灣餐飲企業培養了無數的餐廚精英，現轉任臺北城市科技大學，擔任更上一層樓的餐旅行政工作。一路走來勤勤懇懇，用最專業的態度面對廚之一道。

　　周教授的精進版新書《西餐料理一把罩》是一本非常突出的佳作，不但囊括各種香料、調味料、度量衡介紹，也包括了高湯的基本功，從基礎開始夯實讀者的基底。讓我最稱讚的是他以「套餐」方式來詮釋菜單的管理方法和製作流程。菜單管理是一間餐廳的靈魂，囊括著一間餐廳的中心思想和獨特「菜格」。本書共有十五種不同的套餐，共90道菜品，包括前菜、沙拉、湯品、主菜和甜點，是一本應有盡有的廚藝工具書和實用寶典。

　　讀者可以在品讀與實做之既，領悟周教授套餐菜品搭配的「風格」與「和諧」，不為了創意而創意，在創意中兼顧整體和諧，讓饕客在品嚐每一道菜品之時，更可領悟套餐的美好。這是一本不可多得的餐飲巨作，非常開心有這個機會為本書作序，真摯推薦，願本書成為您廚房中的良「書」益友。

前美國藝術大學鳳凰城分校廚藝管理學院院長
美國廚藝協會廚藝學院資深院士
美國廚藝學院名人殿堂院士
世界廚藝協會A級國際評委

Chef Works

雪沃 質感專業職人制服

台北大直 展示間
營業時間：平日 10:00-17:00
預約電話：02-8502-7092
Email：info@chefworks.com.tw
地址：104 台北市中山區樂群二路101號1樓

台中西屯 展示間
營業時間：平日 10:00-18:00
預約電話：04-2463-1801
Email：info@chefworks.com.tw
地址：台中市西屯區福裕路85號

官網線上購物服務
www.chefworks.com.tw

特別感謝

周慶雄、蘇祖壽、黃正斌、黃維君、符聖憶、郭品岑、杜建文、呂漢強、李旻玲、余紫彤、練旻穎、
陳樂宇、高家偉、邱奕承、陳威銘、蘇禹華、簡上勳、李茂霖、彭啟軒、旭森餐具、
Chef Works 廚師服及我最親愛的家人鼎力相助

推薦序

中國人會做中國菜是本能，但會做西餐就得花上三年四個月才會出師。學西餐真的是那麼難嗎？在 Andy 老師的指導下「絕不難」，而我就是一個活生生的例子。

不敢欺瞞大眾，平常要我上中餐、烘焙等課程，那是易如反掌，但要我做西餐，就真的有點尷尬了。記得兩年前，在好友戴淑貞老師的引薦下，需至台南女子技術學院家政系擔任教授西餐實習的課程。雖然我能以中餐的教學能力當底子，但是中、西餐南轅北轍，並不是我們用中餐烹調的觀念就能應付的。因此，為了不使學生對我的教學感到失望，遂鼓起最大的勇氣去請教 Andy 老師。果真不出我所料，在他熱情、專業、正確的口述指導後，便使我輕易的踏上西餐教學的路程。

學問需要延綿，技能更需要傳承，有如此好手藝、好方法的西餐老師，不叫他教出點撇步怎麼可以？很高興在三藝文化事業有限公司薛先生的邀約下，Andy 老師終於出書了。更興奮的是，我能在這本書上軋上一腳，寫了推薦序。

Andy 老師以 15 年主廚及教學經驗，將所設計的好菜單，搭配簡單的材料與詳細的製作方法，就能讓讀者一目瞭然，輕鬆完成一套套色香味俱全的菜餚，所以除了研讀之外。別忘了，還要享受 DIY 的樂趣喔！

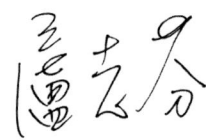

推薦序

從育達高中餐飲管理科創科至今，深深感動於 Andy 老師教學待人的熱誠。常常能在閒聊中，就能激發創意教學，真正落實快樂學習的目標。

Andy 老師所著"西餐料理一把罩"，透過系統教學，深入淺出使熱愛西餐料理的讀者透過專業的食譜，學習，取量及了解色、香、味烹調技藝的核心，書上精美的排盤，餐具的選擇應用。使讀者能學習形、器、美的概念。這是一本好書，極力推薦您!!

育達高中餐飲管理科 科主任

推薦序

老子道德經：第六十章【原文】治大國，若烹小鮮。

以道蒞天下，…治大國，若烹小鮮：通常都認為這句話的言下之義是〝不要隨便翻弄〞的意思。這是始於韓非的解釋。因為「君無為而法無不為；法行而君不必憂，臣不必勞，民但守法，上下無為而天下治。」和老子「無為而治」的政治理念是很相似的。

周主任用檸檬鮭魚捲與奶油芥末醬這一道菜，把〝烹小鮮（魚）〞的哲學發揮的淋漓盡致。三道火候：一煮、二煎、三烤；各有不同的工具、火力、時間與技巧，經歷三道嚴謹的工序處理之後，魚捲體型完整、色澤油潤、肉質挺實。醬汁的調理更是登峰造極，巧妙地運用柳橙與檸檬皮末的苦澀，襯托出芥末子醬的辛味，呼應著遠方紅蕃茄的酸甜；加熱濃縮的鮮奶油與白酒，散發著濃郁的香氣如同母親的慈愛，溫馨包容著所有的辛、酸、苦，鮭魚捲就像那浮雲遊子，歷盡滄桑之後而重返母親懷抱。這只是 Andy 所著〝西餐料理一把罩〞這本書所介紹的母親節套餐系列其中之一，其餘的等有機會品嚐以後再介紹，或者自己買一本回家體驗一下吧！

經國管理暨健康學院 董事長 蔣孝希

推薦序

我吃到 Andy 師傅親手做的菜，不是在餐廳，也不是在學校，而是在攝影棚，錄製我所製作的科普節目『ㄐㄧㄤˋ吃就對了！』。每一次每一集 Andy 師傅在現場展現他的廚藝時，光是用眼睛看、耳朵聽，就讀到了他烹煮出來的香味和美味，更甭說實際品嚐時，那真是全身上下的能量都被啟動了！真的，一點也不誇張，他的廚藝作品正如他的人一樣，充滿活力！

「好吃的食物都不營養、營養的食物都不好吃」？我知道央求一位忠於美味的主廚去替營養飲食背書，是一件很為難的事，但這正是『ㄐㄧㄤˋ吃就對了！』想要扭轉的方向。起初我找上 Andy 師傅一起參與節目製作，是看中他跨足餐飲界和學界，應該能夠融合健康的的概念，在美味與營養之間拿捏得宜；果不其然，一開始錄製前幾集的時候，對於他所開出的食譜受到來自於營養學界的挑戰，感到相當不適應！但後來為了配合少油、少鹽、少糖的節目宗旨，他開始調整食譜的配比。26集節目錄下來，與其說他料理的方式改變了，不如說他的廚藝不斷在精益求精，做出來的每道菜，力求讓人吃得心曠神怡！

以 Andy 師傅對烹調技術的要求態度，我相信他這本『西餐料理一把罩』一定能夠帶給你層次豐富的參考經歷，讓你隨時都能聞得到美味，享受得到健康！

科普影片製作人 袁瑗

Contents

- 8   各種香料介紹
- 12 調味料 / 度量衡介紹
- 13 高湯介紹

15 情人節套餐
Set Menu for Valentines' Day

- 16 鮪魚沙拉附紅椒杏仁醬
  Ahi Tuna Salad with Red Bell Pepper Almond Dressing
- 18 奶油蘆筍湯
  Cream of Asparagus Soup
- 19 鄉村洋芋泥
  Mashed Potato
- 20 迷迭香烤雞附原味肉汁
  Roasted Chicken with Rosemary and Chicken Gravy
- 22 夏威夷核果羊小排
  Hawaii Macadamia Lamb Chop
- 24 皇家巧克力慕斯
  Royal Chocolate Mousse

26 兒童套餐
Set Menu for Children

- 27 酥皮奶油玉米湯
  Puff Pastry on Cream of Corn Soup
- 28 鳳梨芒果蝦仁
  Grass Shrimp Salad with Mango and Pineapple Relish
- 30 漢堡排附薯條及高麗菜沙拉
  Hamburger with French Fries and Cole Slaw
- 32 奶香培根義大利麵
  Spaghetti Carbonara
- 34 焦糖布丁
  Caramel Pudding

36 母親節套餐
Set Menu for Mothers' Day

- 37 餐包
  Bread Roll
- 38 凱撒沙拉與香烤雞胸
  Caesar Salad with Grilled Chicken Breast
- 40 法式洋蔥湯
  French Onion Soup
- 42 黑胡椒牛肉乳酪烘蛋
  Black Pepper Beef and Cheese Frittata
- 44 檸檬鮭魚捲與奶油芥末醬
  Lemon Salmon Roll with Cream Mustard Sauce
- 46 英式麵包布丁
  Butter Bread Pudding

47 父親節套餐
Set Menu for Fathers' Day

- 48 海鮮沙拉盅
  Seafood Cocktail
- 49 義式蔬菜湯
  Minestrone
- 50 碳烤菲力牛排與黑胡椒醬
  Grilled Beef Tenderloin with Black Pepper Sauce
- 52 藍帶雞排附香料蕃茄
  Chicken Cordon Bleu with Bake Tomato
- 54 提拉米蘇
  Tiramisu

56 聖誕節水果套餐
Christmas Fruit Set Menu

- 57 肉桂紅酒梨
  Spiced Pears in Red Wine Sauce
- 58 奶油玉米甜瓜湯
  Toasted Corn and Sweet Squash Soup
- 59 德國聖誕紅酒
  Gluehwein
- 60 橙香夏威夷鮭魚沙拉
  Lomi Lomi Salmon
- 62 香烤果餡火雞捲
  Roasted Turkey Breast with Fruit Stuffing
- 64 酪梨香草雪球
  Snow Ball

65 中國新年套餐一
Chinese New Year Set Menu 1

- 66 焗奶油明蝦生蠔
  Creamed Oyster and Shrimp in Shell
- 68 豬肋排附美式烤肉醬
  Pork Ribs with B B Q Sauce
- 70 水果巧克力火鍋
  Chocolate Fondue with Fresh Fruit
- 72 蘑菇蓮子卡布奇諾湯
  Mushroom and Lotus Seeds Cappuccino Soup

73 中國新年套餐二
Chinese New Year Set Menu 2

- 74 烤鴨香梨華爾道夫沙拉
  Pear Waldorf Salad with Roasted Duck
- 75 青醬
  Pesto Sauce
- 76 義式蕃茄海鮮鍋
  Italian Tomato Seafood Fondue
- 78 無骨牛小排附蘑菇白蘭地醬
  Slices Boneless Beef Short Ribs with Mushroom Brandy Sauce
- 80 金桔奶酪
  Kumquat Panna Cotta

82 早餐套餐
Breakfast Set Menu

- 83 法國吐司
  French Toast
- 84 活力蘿蔔汁與熱巧克力咖啡拿鐵
  Carrot Revitalizer and Hot Chocolate Coffee Latte

85 水果燕麥片
　 Bircher Muesli
86 火腿乳酪綜合蛋捲
　 Ham and Cheese Omelet
88 優格馬鈴薯沙拉鱸魚排
　 Yogurt Potato Salad with Pan fried Sea bass

90 早午餐套餐
Brunch Set Menu

91 美式鬆餅附楓糖醬
　 Pancake with Maple Syrup
92 麵包甜玉米巧達湯
　 Sweet Corn Chowder in French White Bread
94 焗奶油菠菜烘蛋
　 Baked Egg with Creamy Spinach
95 香煎早餐腸、培根及火腿
　 Fried Breakfast Sausage, Bacon and Roll Ham
96 楓香奶昔
　 Maple Leaf Frappe

97 午茶休閒餐
Afternoon Tea Menu

98 鮪魚酪梨沙拉
　 Tuno and Avocado Salad
100 乳酪雞肉派
　 Chicken Pot Pie
101 英式比士吉
　 Scottish Scones
102 火腿蒜苗塔
　 Ham and Leek Quiche
104 蘋果馬芬蛋糕
　 Apple Muffin

105 DVD 風味休閒餐
Great DVD Set Menu

106 蛤蜊巧達湯
　 Clam Chowder
108 乳酪烤淡菜
　 Cheesy Grilled Mussels
109 海鮮披薩
　 Seafood Pizza
110 鮮蝦酪梨三明治
　 Grass Shrimp and Avocado Sandwich
112 香酥雞肉條與炸起司條附塔塔醬
　 Chicken Fingers and Cheese Sticks with Tartar Sauce
114 香蕉奶昔
　 Banana Ice Cream Shake

115 三明治套餐
Sandwich Set Menu

116 奶油白花菜湯
　 Cream of Cauliflower Soup
117 鮪魚潛水艇三明治
　 Tuna Submarine
118 總匯三明治
　 Club Sandwich
119 蛋沙拉三明治
　 Egg Salad Sandwich
120 夏日水果沙拉三明治
　 Summer Fruit Salad Sandwich

121 家庭宴會餐
Menu for Family Party

122 芒果水果汁
　 Mango Fruit Cooler
123 綜合蘑菇沙拉
　 Mixed Mushroom Salad
124 匈牙利牛肉湯
　 Beef Goulash Soup
125 焗海鮮飯
　 Gratin Seafood Rice
126 義大利肉醬麵
　 Spaghetti Bolognaise
128 咖啡慕斯
　 Home Made Coffee Mousse

129 全素套餐
Vegetarian Set Menu

130 翠玉花菜沙拉
　 Broccoli and Cauliflower Salad
131 奶油青豆仁湯
　 Cream of Green Pea Soup
132 蔬菜千層麵
　 Vegetable Lasagna
134 英式草莓鬆餅
　 Strawberry Shortcake

135 烤肉餐
BBQ Menu

136 蒜味香料干貝草蝦串
　 Herb Garlic Scallop and Prawn Skewer
138 墨西哥烤牛排附酪梨醬
　 Mexican Steak with Avocado Salsa
140 烤蔬菜串
　 Grill Vegetable Kebabs
141 碳烤原味玉米
　 Corn on the-Cob
142 椰香烤鳳梨
　 Pina Colada Pineapple

143 附錄

※ 本書的菜式、均為四人份

◆ 辣椒 ◆
Chili Pepper, Cayenne Pepper

辛辣香料，乾製後磨成粉狀，適用於墨西哥、東南亞、泰國、馬來西亞等地區較辛辣之料理。著名的辣椒水 Tabasco 就是墨西哥的辣椒所製造的。

◆ 香草 ◆
Vanilla

如豆莢的果實，於未成熟時採收，主要用於甜點 Dessert。除有天然原狀香草外，尚有磨成粉或香精及合成的香草精（較便宜），但以天然者為佳。

◆ 月桂葉 ◆
Bay leaf

主要為增加液體烹調的風味，適用於湯、醬汁，燴煮類或水煮類。大部分以乾燥的居多。

◆ 百里香 ◆
Thyme

具有強濃辛烈的風味可去除異味，適合肉類、海鮮、醬汁，家禽的食物。

◆ 匈牙利紅椒粉 ◆
Paprika

大部分產品沒有辣，具香甜味，顏色為桃紅色，可做著色裝飾。原產地為匈牙利，適用於匈牙利燴肉 Goulash、飯類及燴菜類。

◆ 紅酒醋 ◆
Red Wine Vinegar

為紅酒發酵釀造而成，亦有白酒醋之分。適用於醬汁、沙拉醬的調味，也適合加入新鮮香料醃製成香料酒醋。

◆ 荳蔻及荳蔻皮 ◆
Nutmeg and Mace

為桃狀果實的種子，去皮稱荳蔻，其合皮稱荳蔻皮，味道相近，適用於洋芋泥、菠菜等蔬菜類及燴菜類。

◆ 蝦夷蔥 ◆
Chive

屬蔥類植物，為綠色，呈細長葉子狀，有著色作用，也有蔥的香味。適合裝飾在湯及沙拉，若不易購買，亦可用青蔥綠色的部分代替。

◆ 薄荷葉 ◆
Mint

清涼味適合與水果、沙拉、甜點搭配及裝飾用。

◆ 黑橄欖 ◆
Black Olive

為橄欖樹之熟果實醃製而成；未成熟則為綠色。適用於沙拉、醬汁等配菜。

◆ 丁香 ◆
Cloves

可增加火腿或豬肉的風味。分成和月桂葉一起使用的水煮類及用在烘焙的粉狀物。

◆ 鼠尾草 ◆
Sage

適用於香腸肉餡、家禽類，魚或豬肉類的填塞餡 Stuffing，可去除異味增加香味。

◆ 茵陳蒿 ◆
Tarragon

為法國人的最愛,可去除魚、肉類之異味。常漬於醋中,稱茵陳蒿醋 Tarragon Vinegar,適用於貝爾納斯醬汁 Bearnaise Sauce 中的香料,可與牛排、蛋類搭配。

◆ 酸豆 ◆
Caper

為續隨子樹的花蕾,一般皆浸漬於醋中,可增加沙拉醬的風味,多用做海鮮的配料及燻鮭魚佐料。

◆ 羅勒 ◆
Basil

味道與九層塔相似,但比九層塔溫和,適用於湯、披薩醬汁、義大利麵等,是義大利人最愛的香料。常與蕃茄料理一起搭配或做成青醬 Pesto,可搭配海鮮類、麵類 Pasta、肉類等。

◆ 法式芥末 ◆
French Mustard

辛辣味,黃色,適合用於紅肉類、熱狗等,也有著色作用,可使醬汁或燴煮菜類變黃。

◆ 奧勒岡(花薄荷)◆
Oregano

為義大利菜主要的香料,適合用在魚類,蕃茄,肉類等菜的調味。

◆ 洋香菜,巴西利 ◆
Parsley

可增進湯、醬汁風味,新鮮的則可做裝飾。

◆ 迷迭香 ◆
Rosemary

適合燴煮類、爐烤類,其特殊香味可蓋過肉的異味,可漬於橄欖油中來醃漬羊排,魚排等。

◆ 香菜 ◆
Coriander, Chinese Parsley

具有特殊辛烈風味,適合墨西哥、義大利及中國料理。種子稱 Coriander Seed,也常磨成粉來使用。

◆ 肉桂 ◆
Cinnamon

為條狀或磨成粉的。主要用於甜點,如蘋果派或撒在 Cappuccino 咖啡的泡沫上。

◆ 蒜頭 ◆
Garlic

辛辣香味可增進食慾,適合與肉類、海鮮、蔬菜等類做搭配。

◆ 胡椒 ◆
Papper Cone

胡椒子未成熟時即採收,乾燥而成黑胡椒。味道強烈,適合紅肉類的調味,成熟後的胡椒子去皮而成白胡椒,其味道較不強烈,適合白肉類及魚類的調味。

◆ 蒔蘿 ◆
Dill

適用於醃漬類及沙拉的調味,如黃瓜沙拉等或魚類、湯類、燴煮類等料理。

調味高湯介紹

決定一道菜的成功與否,除考慮其烹調方式、營養成分、整體外觀外,最重要的特色就是風味(Flavor)。影響風味的主要關鍵在於如何使用西餐的調味料及正確的調味。

選用良好的高湯 Stock 代替味精。

1. 雞高湯　Chicken Stock
2. 魚高湯　Fish Stock
3. 牛高湯　Beef Stock
4. 香菇高湯　Mushroom Vegetable Stock

正確調味 Seasoning。

西餐的調味 Seasoning 主要指鹽及胡椒。(海鮮類可加上檸檬汁。)
Seasoning means salt and pepper. (Could add lemon juice with seafood.)

選擇天然的蔬菜及香料增加其風味。

1. 調味蔬菜 Mirepoix:包括了洋蔥50%(或洋蔥25%、蒜25%)、紅蘿蔔25%、西芹25%。
2. 香料束 Bouquet garni:將西芹、百里香、洋香菜等香料細綁成束。
3. 香料包 Spicy Bag:用紗布將綜合香料,如胡椒粒、月桂葉、百里香等包在一起。

---

1. Mirepoix is a combination of rough-cut vegetable:50 % onions (or 25 % onions, 25 % leek), 25 % carrots, 25 % celeries.
2. Bundle of herbs: Tie assorted herbs to be a bundle, such as celeries, thyme bay leaf and parsley and so on.
3. Spices bag: Is any mixture of herbs and spices tied in a square of cheeseclose.

度量衡及溫度的換算

重量的換算
1 公斤 = 1000 公克
1 公斤 = 2.2 磅
1 盎司 = 28.4 公克
1 磅 = 454 公克
1 磅 = 16 盎司
1 公斤 = 0.6 台斤
1 台斤 = 16 兩

容量的換算
1 茶匙 = 5 cc 公撮
1 湯匙 = 15 cc 公撮
1 杯 = 250 cc 公撮
1 品脫 = 500 cc 公撮
1 夸脫 = 1000 cc 公撮(1 liter)
1 加侖 = 4000 cc 公撮(4 liter)

雞高湯（2 公升）

材料 / Ingredients

2 1/2 公升	水	water	調味蔬菜 Ingredients	60 公克	紅蘿蔔	carrot
1 公斤	雞骨	chicken bone		60 公克	西芹（清洗切小丁）	celeries
				125 公克	洋蔥	onion

香料包 / Spices bag

5g	迷迭香	rosemary
5g	百里香	thyme
1 片	月桂葉	pc bay leaf
1 枝	巴西利梗	stalks of parsley
5 顆	白胡椒粒	white pepper corns

作法 / Method

1. 燙煮：將水加入湯鍋煮至沸騰，加入雞骨，用木匙攪拌均勻。（主要去除多餘的血水及油脂）
2. 過濾：將雞骨倒入漏水盆，使雞骨濾乾。
3. 將雞骨再放入湯鍋，加入冷水煮至沸騰，去泡沫及殘渣。
4. 將火轉小，加入調味蔬菜及香料包，使高湯濃縮。
5. 不斷去除浮在高湯的表面泡沫及殘渣，用慢火煮至 1 1/2 小時。
6. 最後過濾即可。

---

1. Blanch: put chicken bones into the boiled water, stir it with wooden spoon (it is for wiping blood and fat), and strain the chicken bones.
2. Put bone in a pot with cold water and bring to a simmer and skin. Add mirepoix and spicy bag and return to a simmer.
3. Simmer for 1 1/2 hour, skimming occasionally.
4. Strain and degrease.

魚高湯（2 公升） Fish Stock (2 litre)

材料 / Ingredients

2 1/2 公升	水	litre of water
1 公斤	魚骨	kg fish bones
120 公克	洋蔥	onion
60 公克	蒜（只用白色部分）	leek (only use the white part)
1/2 顆	檸檬汁	lemon juice
120cc	白酒	white wine
15 公克	奶油	nutter

香料包 / Spices bag

2 枝	巴西利梗	stalks of parsley
1 片	月桂葉	pc bay leaf
5 顆	白胡椒粒	white pepper corns

作法 / Method

1. 若魚骨不新鮮或有血水時需清洗魚骨。
2. 將洋蔥及蒜切片用奶油將洋蔥及蒜炒香，不上色。
3. 再將魚骨及炒過的蔬菜放入高湯鍋，加入檸檬汁。
4. 加入冷水，煮至沸騰，轉小火。
5. 去除泡沫加入香料及白酒。
6. 慢煮 20 ~ 25 分鐘，慢煮過程中不斷去除泡沫及殘渣。

1. Wash fish bones if it is not fresh or with blood on it.
2. Cut onions and garlic to be slices then pan-fry with butter.
3. Then put the fish bones and fried vegetables into the pot, add some lemon juice into it.
4. Add cold water into the pot then bring to a simmer.
5. And skim, add herbs and white wine into the pot.
6. Simmer for 20~25 minute, strain and degrease.

褐色牛高湯（2 公升）Brown Beef Stock（2 litre）

材料 Ingredients

3公升	水		litre of Water
1公斤	牛骨（切成小塊）		kg of beef bones (cut to be small sizes)
120g	洋蔥		onions
60g	紅蘿蔔		carrots
60g	西芹		celeries
100g	蕃茄糊		tomato paste

香料包 Spices bag

5g	迷迭香	rosemary
5g	百里香	thyme
3片	月桂葉	pc bay leaf
2枝	巴西利梗	stalks of parsley
10顆	黑胡椒粒	black pepper corns

作法 Method

1. 將牛骨放入烤盤，進烤箱 200℃ 烤。
2. 當牛骨變淺褐色時，加入調味蔬菜及蕃茄糊。蕃茄糊均勻塗在牛骨上，再入烤箱，烤至深褐色（約 15～20 分鐘）。
3. 再將深褐色的牛骨放置高湯鍋中。
4. 去殘渣：用水或紅酒，將烤盤上的殘渣去除，再將其液體倒入高湯中。
5. 加入冷水至高湯鍋中，再加香料袋。
6. 煮至沸騰後小火慢煮 6～8 小時，過程不斷去除浮油及泡沫。
7. 過濾即可。

---

1. Brown bone in a hot oven 200℃.
2. Add mirepoix and tomato paste while beef bones become light brown. Spread tomato paste on bones then put it into oven and bake until the colour becomes dark brown (about 15 ~ 20 minutes).
3. Then put dark brown beef bones into the pot.
4. Deglaze: To reject dregs on the baking tray with water or red wine, then pour the liquid into the pot.
5. Add cold water and spice bag.
6. Bring to a boil and skim and simmer for 6 ~ 8 hours, skimming occasionally.
7. Strain and degrease.

香菇蔬菜高湯（2 公升）Mushroom Vegetable stock（2 litre）

材料 Ingredients

3公升	水	Water
100g	乾香菇	dry mushroom
300g	紅蘿蔔塊	carrot cubes
300g	西芹塊	celeries cubes
2顆	蕃茄塊	Tomato cubs
1片	月桂葉	Bay leaf
1支	巴西利梗	Stalk of parsley

作法 Method

1. 將所有材料大火煮至沸騰，轉小火濃縮 60 分鐘，過濾即可。

1. Put all Ingredient in to pot with cold water and bring to a simmer and skin for 60 min. After strain and degrease.

Set Menu for Valentine's Day
情人節 套餐

Ahi Tuna Salad with Red Bell Pepper Almond Dressing
鮪魚沙拉附紅椒杏仁醬

Cream of Asparagus Soup
奶油蘆筍湯

Mashed Potato
鄉村洋芋泥

Roasted Chicken with Rosemary and Chicken Gravy
迷迭香烤雞附原味肉汁

Hawaii Macadamia Lamb Chop
夏威夷核果羊小排

Royal Chocolate Mousse
皇家巧克力慕斯

在情人節這個浪漫的日子裡，四處可見鮮花、巧克力及各種禮物，這不僅是情侶的專利，也是屬於夫妻之間的浪漫，藉著這天來表達彼此心中的愛意，一頓美味的餐點再襯以絕佳的氣氛，更為這個節日加分不少。這套情人節的專屬套餐包含了入口即化的鮮鮪魚沙拉搭配杏仁果香的紅甜椒醬汁，是一道加州式的熱前菜，香醇濃郁的奶油蘆筍湯，主菜則為脆皮多汁的迷迭香烤雞及夏威夷核果羊小排，餐後再搭配濃情的巧克力慕斯。整份菜單若能搭配香檳及佐餐紅白酒，再加上玫瑰與蠟燭，會使這套菜單更出色，更增添羅曼蒂克的氣氛。

Flowers, chocolates and all kinds of gifts are everywhere in the romantic day such as Valentines' Day. It is not only lovers' patent but also belonged to couples' romance. This set menu which belongs to Valentines' Day contains ahi tuna salad with almond and red pepper sauce, which is a hot Californian appetizer, soup is creamy asparagus soup, the main course is crispy roasted chicken with rosemary, and the dessert is rich chocolate mousse. If the entire menu with champagne and table wine, and decorate the environment with rose and candles would make this dish better and more romantic.

鮪魚沙拉附紅椒杏仁醬
Ahi Tuna Salad with Red Bell Pepper Almond Dressing

難易度 ★★★　　　時間 45 分

情人節套餐

材料 Ingredients

40g	杏仁片	almond slices	
1 顆	紅甜椒	red bell pepper	
5g	蒜頭碎	chop garlic	
1 顆	蛋黃	egg yolk	
170cc	橄欖油	olive oil	
10cc	紅酒醋	red wine vinegar	
250g	鮪魚排	切成兩片約 1.2 公分厚 tuna fillet cut to be two 1.2 cm thick parts	
適量	黑胡椒碎、鹽、白胡椒粉 T.T. black pepper, salt, white pepper		

A
- 100g 苜蓿芽 alfa alfa sprout
- 100g 豌豆苗 pea shoots
- 50g 紅蘿蔔絲 carrot, julienne
- 1/4 顆 黃甜椒絲 yellow bell pepper, julienne
- 150g 高麗菜絲 cabbage, julienne

B
- 30g 檸檬汁 lemon juice
- 90g 橄欖油 olive oil

作法 Method

1. 烤箱預溫 180 度後，將杏仁片放入烤箱中，烤至呈金黃色，待冷，切碎備用。
2. 紅甜椒碳烤或放入烤箱烤至焦黃上色，去皮去籽，再用攪拌機打成泥狀。
3. 蛋黃加少許的溫水及紅酒醋，打散後加入大蒜碎，徐徐加入橄欖油以打蛋器打成稠狀，再加入杏仁片及紅椒泥、鹽、辣椒粉調味即為紅椒杏仁醬。
4. 將材料 A 的綜合生菜洗淨過冷水，過濾備用。
5. 將材料 B 中的檸檬汁及橄欖油混合，打散成稠狀檸檬醬汁並用鹽、胡椒調味。
6. 鮪魚用黑胡椒碎及鹽調味，用中大火煎至兩面 5～7 分熟後切斜片（一塊鮪魚排約 8 小片）。
7. 鮪魚上盤附上綜合生菜，再淋上紅椒杏仁醬及檸檬醬汁即可。

---

1. Toasted almond slices until golden brown and chopped.
2. To grill or bake red bell peppers until the colour turned to be brown and burned, take away the skin and seeds then blend to be mashed.
3. To scatter few warm water and red wine vinegar with egg yolk, then add chopped garlic into it, and then add olive oil and mix well, then add chopped almond, mashed red bell pepper. Salt and chili powder to be red bell pepper dressing.
4. Clean the assorted lettuces (ingredients A) then strain it.
5. Mix ingredient B equally to be lemon dressing and season with salt and pepper.
6. Season tuna with chopped black pepper and salt, then pan-fry the both sides of tuna until medium-well then sliced.
7. Put sliced tuna with assorted lettuce and then pour red pepper almond dressing and lemon dressing in a plate.

奶油蘆筍湯
Cream of Asparagus Soup

難易度 ★　　　　時間 30 分

材料 Ingredients

600g	綠蘆筍（小）	green asparagus (small)
1公升	雞高湯	litre chicken stock
75g	奶油	butter
75g	低筋麵粉	flour
170cc	無糖鮮奶油	fresh cream
適量	鹽、胡椒	T.T. salt, pepper

A	1/2 顆	洋蔥	onion
	1 支	西芹	stalk of celery
	1 支	青蒜	leek
B	1 支	巴西利梗	stalk of parsley
	1 片	月桂葉	bay leaf
	2.5g	百里香	thyme

作法 Method

1. 綠蘆筍留尖頭部分約 1 杯，以水燙熟後沖冷水備用。
2. 將其餘蘆筍切小丁，材料 A 的調味蔬菜亦切丁備用。
3. 用奶油將材料 A 及蘆筍拌炒約 1 分鐘至其散發出香氣，續放入麵粉炒成白色麵糊，再加入雞高湯攪拌至無顆粒狀，煮開後轉小火，再加入材料 B，以小火煮 20 分鐘，過濾。
4. 加入鮮奶油煮至沸騰，以鹽、胡椒調味，最後加入裝飾用蘆筍即可。

1. Keep the top part of green asparagus for about 1 cup and blanch.
2. Cut the rest of asparagus to be dice, cut the mirepoix (ingredients A) to be dice.
3. Sweat ingredients A and asparagus with butter for about 1 minutes then add flour to be white flour paste (roux), then add chicken stock to dissolve roux. Bring to simmer then add ingredients B and simmer for about 20 minutes then strain.
4. Blend in cream and boil until done. Season to taste and garlish with asparagus.

鄉村洋芋泥
Mashed Potato

難易度 ★　　　　　時間 20 分

材料 / Ingredients

600g	洋芋（約 2 - 3 顆）	potatoes (about 2 - 3 each)
2 顆	蛋黃	egg yolk
30g	奶油	butter
60g	鮮奶油	fresh cream
適量	鹽、胡椒、豆蔻粉	T.T. salt, pepper and nutmeg.

作法 / Method

1. 洋芋去皮切大塊，用水煮約 15 分鐘至熟，過濾，再壓成泥。
2. 拌入奶油及鮮奶油，以小火煮熱，再拌入蛋黃及鹽、胡椒，豆蔻粉調味即可。

---

1. Peeled potatoes and cut them to be big cubes, boiled for about 15 minutes, strain them and then press to be mashed.
2. Mix up with butter and fresh cream, and keep cooking then mix up with egg yolk, salt, pepper and nutmeg powder to taste.

迷迭香烤雞附原味肉汁
Roasted Chicken with Rosemary and Chicken Gravy

難易度 ★★★　　　時間 80 分

材料 Ingredients

800g	全雞（2隻）	whole chicken (2 chickens)
3 顆	蒜頭	garlic
50g	奶油	butter
20g	匈牙利紅椒粉	paprika powder
10g	迷迭香	rosemary
適量	鹽、胡椒	T.T. salt and pepper
2 條	棉繩	cotton rope

調味蔬菜 Mirepoix

1 顆	洋蔥	onion
2 支	西芹	stalk of celeries
1 條	紅蘿蔔	carrot

情人節套餐

作法 Method

1. 烤箱預熱 180℃。
2. 將調味蔬菜切約 2 公分立方，加入蒜頭及迷迭香混合均勻。
3. 全雞洗淨後，用鹽、胡椒、匈牙利紅椒粉調味，再把部份的調味蔬菜塞入雞肚中以棉繩綁緊。
4. 將剩餘的調味蔬菜墊在烤盤底部，把全雞放入烤箱，每 15 分鐘翻一次，並淋上滴油，烤至整隻雞呈金黃色約 1 小時至熟。
5. 雞烤好後，將其背骨及胸肋骨取出，再將全雞一開為二，腿骨可留下（一人份半隻）。
6. 原味肉汁：將烤盤上的調味蔬菜、取下的雞骨及烤雞留下的油脂倒入鍋中，加入約 500cc 的雞高湯，濃縮，15 分鐘後撈去浮油並過濾，再以鹽、胡椒調味即可。
7. 上盤時淋上原味肉汁，並附上鄉村洋芋泥。

---

1. Preheat oven to 180℃.
2. Cut mirepoix to be about 2 cm cubes then mix with garlic and rosemary.
3. Clean the whole chickens and season with salt, pepper and paprika powder, and then stuff some of mirepoix into chicken and truss.
4. Spread the rest of mirepoix on the bottom of the baking tray, put the whole chicken into the oven to roast, turn it every 15 minutes until golden brown for about 1 hour and well down.
5. Boneless chicken and cut it to be two parts, keep the bones of legs (half chicken each person).
6. Chicken Gravy: put the mirepoix and chicken bones which is from baked chicken into a pot, add approximate 500cc of chicken stock into it, bring to simmer for 15 minutes, skimming and season with salt and pepper.
7. Pour the chicken gravy with chicken in a plate, and with mashed potato on the side.

夏威夷核果羊小排
Hawaii Macadamia Lamb Chop

難易度 ★★★　　　　時間 80 分

材料 Ingredients

12支	羊小排	lamb cutlets	① 50g	夏威夷果	macadamia
			20g	黃芥末醬	yellow mustard
適量	鹽、胡椒	T.T. salt and pepper	10g	蒜頭	garlic
			5cc	白蘭地	brandy
			2g	迷迭香	rosemary
			2g	薄荷葉	mint leaves

配菜 Mirepoix

4 小朵　青花菜　　　　　pieces of broccoli
4 支　　玉米筍　　　　　baby corn
1/2 顆（切 4 片）紅甜椒　red bell pepper (cut into 4 slices)
※ 青花菜，玉米筍和紅甜椒燙煮沖冷水備用
※ Blanch broccoli, baby corn and red bell pepper, then rinse them with cold water and set aside

4 顆　　洋芋 (350g)　　　potatoes
1 支　　蔥（切蔥花）　　spring onion(chopped)
20g　　奶油　　　　　　butter
20g　　酸奶油　　　　　sour cream

作法 Method

1. 烤箱預熱至 180℃；蒜頭及巴西利切碎；①料拌勻備用。
2. 將洋芋用鋁箔紙包起，入烤箱 180℃約 1 小時，烤洋芋用小刀劃十字，由內往外壓成洋芋花再加上蔥花、酸奶油及奶油。
3. 將配菜青花菜、玉米筍和紅甜椒用奶油炒香。
4. 羊小排先用鹽、胡椒調味，入鍋煎 3 分鐘後，入烤盤內，再以①料鋪於羊小排上，入烤箱烤約 10 分鐘至 8 分熟即可。

---

1. PPreheat oven to 180 ℃. Chop up the garlic and parsley stalk.
   Mix well ingredients ①, then set aside. .
2. Pack the potatoes with foil, then put into the oven at 180℃ for about 1 hour. Slice a cross into the top of each potato with a knife, then push in on the ends of each potato, and garnish with chopped spring onion, sour cream, and cream.
3. Saute the mirepoix (broccoli, baby corn, and red bell pepper) with butter until fragrant.
4. Season lamb cutlets with salt and pepper, and pan-fry for three minutes. Then put them in baking tray, top with ingredients ① and put in oven for 10 minutes to a medium-well.

皇家巧克力慕斯
Royal Chocolate Mousse

難易度 ★   時間 30 分

材料 Ingredients

	150g	苦甜巧克力	dark chocolate
	250cc	無糖鮮奶油	fresh cream
	30cc	櫻桃白蘭地或白蘭地	cherry brandy or congac
	5cc	香草精	vanilla extract
		草莓、奇異果及薄荷葉適量（洗淨）	strawberry, kiwifruit and mint leaf (cleaned)
A	3 顆	蛋黃	egg yolks
	20g	細糖	sugar
B	3 顆	蛋白	egg white
	20g	細糖	sugar
C	45cc	滾水	boiled water
	5g	咖啡粉	instant coffee powder

情人節套餐

作法 Method

1. 將巧克力切碎，隔水加熱融化。
2. 用打蛋器將材料 A 打至微發，材料 B 的蛋白亦打發備用，材料 C 攪拌勻成濃縮咖啡。
3. 將鮮奶油打發，預留 1/3 杯裝飾用。
4. 在融化的巧克力中，依序拌入蛋黃、蛋白、濃縮咖啡及打發鮮奶油拌勻，再加入櫻桃白蘭地及香草精攪拌均勻。
5. 將巧克力慕斯倒入香檳杯中，放入冰箱以 5℃ 冷藏至巧克力慕斯凝固，再擠上鮮奶油並飾以草莓、奇異果及薄荷葉。

---

1. Chop chocolate and melt it over hot water.
2. Whip egg yolk with sugar until blended, whip egg whites to stiff peaks with sugar, and mix up the C to be concentrated coffee.
3. Whip fresh cream to be whipped and keep 1/3 of it to be garnish.
4. Mix up equally the stuff (step 2 and 3) in the melted chocolate then add cherry brandy and vanilla extract to mix up.
5. Pour chocolate mouse into a champagne glass and into fridge until done, and then add fresh cream and garnish with strawberry, kiwifruit and mint leaf.

25

Set Menu for Children 兒童 套餐

Puff Pastry on Cream of Corn Soup
酥皮奶油玉米湯

Grass Shrimp Salad with Mango and Pineapple Relish
鳳梨芒果蝦仁

Hamburger with French Fries and Cole Slaw
漢堡排附薯條及高麗菜沙拉

Spaghetti Carbonara
奶香培根義大利麵

Caramel Pudding
焦糖布丁

還記得兒子承威四歲生日時，我準備了材料，到學校與他和及其他同學，合做了一個黑森林生日蛋糕，小朋友那種愉悅的心情及親情的流露，至今仍難以忘懷。學校現在都很注重親子間的互動，你是否曾為了園遊會或生日派對而煩惱呢？給孩子一個美好的童年回憶吧！

這是一套從三歲到十六歲大小朋友都喜歡的菜單，花點時間，也為你的小朋友作套美味營養的套餐吧！

I still remember the party of my son, Chris, whose four-year-old birthday, I prepared some ingredients and make a black forest cake in his school with his schoolmates, the joy and the happiness of sensibility is still recalled in my mind. Teachers nowadays pay more attention to the interaction between kids and parents. Have you ever been bothered by garden parties or birthday parties? Try your best to give your kids a wonderful childhood memories.

酥皮奶油玉米湯
Puff Pastry on Cream of Corn Soup

難易度 ★　　　　　時間 40 分

材料 Ingredients

	份量	材料	Ingredients
	4 張	酥皮（15cm×15cm）	pc puff pastry sheet
	1 顆	蛋	egg
A	25g	奶油	butter
	25g	低筋麵粉	cake flour
B	310cc	牛奶	milk
	500cc	雞高湯	chicken stock
C	250g	玉米醬	corn sauce
	125g	玉米粒	sweet corn
D	1 片	月桂葉	bay leaf
	適量	鹽、胡椒	T.T. salt, pepper

作法 Method

1. 烤箱預溫 220℃。
2. 將材料 A 炒成麵糊，加入材料 B 攪拌成無顆粒狀。
3. 放入材料 C 及材料 D 以小火煮 15 分鐘。
4. 將湯裝入湯碗再蓋上酥皮，擦上蛋液。
5. 入烤箱以 220℃ 烤約 7 分鐘，至呈金黃色即可。

1. Preheat oven to 220℃.
2. Make roux paste with ingredients A, add ingredients B and mix up to dissolve roux.
3. Add ingredients C and D and simmer for 15 minutes.
4. Pour the soup into a soup bowl then cover with the puff pastry sheet, then brush egg wash.
5. Bake at 220℃ in oven for about 7 minutes until the golden brown colour.

鳳梨芒果蝦仁
Grass Shrimp Salad with Mango and Pineapple Relish

難易度 ★　　　　　時間 25 分

材料 Ingredients

1 顆	大芒果（切 0.5 公分立方丁 1 杯，其餘切塊）	mango (cut to be 0.5 cm dice for 1 cup, cut the rest to be big cubes)
1/4 顆	鳳梨（切 0.5 公分立方丁 1 杯，亦可用罐頭鳳梨片）	pineapple (cut to be 0.5 cm dice for 1 cup, could use canned pineapple slices instead of fresh pineapple)
30cc	檸檬汁	lemon juice
10g	香菜碎	chopped coriander
1/2 顆	紅甜椒，切絲	red bell pepper, julienne
30g	美乃滋	mayonnaise
20 隻	草蝦（燙熟，去皮，冷藏備用）	grass shrimps (boiled, peeled, chilled)
2 支	青蔥碎	green onion(chopped)
10g	碎薑	ginger,chop
2.5g	鹽	salt

A
125g	洋蔥塊	onion cubs
120g	西芹塊	celery cubs
90cc	白酒	white wine
1 片	月桂葉	bay leaf

作法 Method

1. 將帶殼草蝦剔去腸泥，可用牙籤在草蝦背部的 2～3 節挑出腸泥。
2. 簡易高湯（Court Bouillon）：將 1.5 公升的水放入燴煮鍋中，加入材料 A，煮開約 10～15 分鐘即成。加入草蝦燙至熟，撈起置於冰水中冷卻，去除頭尾、蝦殼後冷藏備用。
3. 將芒果塊用果汁機打成泥。
4. 將一杯芒果丁及鳳梨丁拌入檸檬汁及香菜碎，加芒果泥及少許的鹽、胡椒調味拌勻為芒果鳳梨醬。
5. 將美乃滋、青蔥碎、薑攪拌均勻，再拌入草蝦及紅甜椒丁。
6. 裝入沙拉碗，加上芒果鳳梨醬，再用紅甜椒絲裝飾即可。

---

1. Take the vein out of the grass shrimp. You can use a toothpick to take out the vein particularly at the 2~3 section of the shell-on grass shrimp.
2. Court Bouillon: put 1.5 litre of water in a pot with (ingredients B) and simmer them for 25 minutes. Add shell-on Giant grass shrimp to the pot and let it boil. When the shrimp is cooked, take them out rinse under iced water. Remove the head, tail, and outer shell of the shrimp and store them in freezer.
3. Puree mango big cube by blender.
4. Mix up a cup of mango dice and pineapple dices with lemon juice and chopped coriander, and add mango puree season with salt and pepper and mix up to be mango pineapple relish.
5. Mix up mayonnaise, chopped green onion and ginger equally then mix with grass shrimps.
6. Put shrimp salad into salad bowl then add mango pineapple relish, and then garnish with red bell pepper julienne.

漢堡排附薯條及高麗菜沙拉
Hamburger with French Fries and Cole Slaw

難易度 ★★★　　　　時間 60 分

高麗菜沙拉 材料
Ingredients of coleslaw

300g	高麗菜絲	cabbage, julienne
80g	紅蘿蔔絲	carrot, julienne
45g	美乃滋	mayonnaise
15g	糖	sugar
15cc	白醋	white vinegar
適量	鹽、胡椒	T.T. salt, pepper

高麗菜沙拉 作法
Method of coleslaw

1. 將高麗菜及紅蘿蔔加少許鹽，用手抓出水分，以冷開水沖洗鹽份再將多餘的水分去除濾乾。
2. 拌入美乃滋、糖及醋，混合均勻後再調味即可。

1. Add little salt into the cabbage and carrot and grab it 1 min. strain out water, then wash it and strain again.
2. Add mayonnaise, sugar and vinegar, mix it up equally then season.

兒童套餐

漢堡 材料 Ingredients for hamburger

400g	牛絞肉（肥肉20％，瘦肉80％）	minced beef (contains 20% fat and 80% meat)
2 片	吐司	pieces of toast
1 顆	蛋	egg
80g	洋蔥碎（奶油炒香備用）	chopped onion (saut?with butter)
60g	蔬菜油	vegetable oil
適量	鹽、黑胡椒、巴西利碎	T.T. salt, black pepper and parsley
4 片	蕃茄片	pieces to tomato
4 片	起司片	slice of cheese
8 片	生菜葉	pieces of lettuce
4 個	漢堡麵包	burger buns
400g	冷凍薯條	French fries

漢堡 作法 Method of burger

1. 將吐司用水泡開，搗成泥狀加入絞肉中，再混合蛋、洋蔥碎、巴西利、鹽、胡椒及油，攪拌均勻。
2. 做成每顆約120g的肉球後，壓扁成牛排形狀，整型時手上可沾點油比較不黏手。
3. 將漢堡排煎到雙面上色，再放入烤箱，以180℃烤至5分鐘，再加入起司片直到起司融化。
4. 準備漢堡麵包，對切，塗上奶油，兩面烤上色。
5. 下層漢堡麵包上加生菜、蕃茄片，再放上起司漢堡排，最後蓋上上層漢堡麵包。
6. 以190℃熱油炸薯條3分鐘至呈金黃色後撈起，以鹽、胡椒調味。
7. 附上高麗菜沙拉及薯條。

---

1. Soaked toast with water then mashed and mix into minced beef, egg, onion, parsley, salt, pepper and oil.
2. Rub it to be several 120g meat balls then press to be steak shape.
3. Pan-fry burger until both sides are golden brown colour then put in oven at 180 ℃ for 5 minutes, then add cheese slices until melted.
4. Prepare bun for burger and cut the half spread butter and toast of the bread.
5. Put lettuce and tomato slices on the below bun, put cheese burger and then cover with another burger bun.
6. Deep-fry French fries at 190℃ for 3 minutes until golden brown and season.
7. The side dishes are coleslaw and French fries.

奶香培根義大利麵
Spaghetti Carbonara

難易度 ★　　　　時間 30 分

材料 Ingredients

400g	義大利麵	spaghetti
500cc	牛奶	milk
100g	培根	bacon
300cc	鮮奶油	fresh cream
100g	新鮮洋菇	fresh mushrooms
15g	蒜頭	garlic
1/2 顆	洋蔥	onion
4 顆	蛋黃	egg yolks
適量	鹽 黑胡椒 巴西利碎 雞高湯	
T.T.	salt, black pepper, parsley (chopped), chicken stock	

作法 Method

1. 將麵條燙煮約 10 分鐘至 8 分熟，沖冷水、瀝乾；洋蔥、蒜頭切碎；洋菇切片備用。
2. 鍋熱入奶油炒香培根後，入洋菇及洋蔥拌炒均勻。
3. 續入麵條、150cc 鮮奶油、牛奶及高湯，濃縮至湯汁剩一半，再入蛋黃及 150cc 鮮奶油，小火濃縮收汁後，最後入加鹽、胡椒 (5:1) 調味，撒上巴西利碎即可。

---

1. Blanch the spaghetti for about 10 minutes to a medium-well. Then rinse with cold water and strain. Chop up onion and garlic. Slice the mushrooms and set aside.
2. Melt the butter in a frying pan and fry the bacon until fragrant. Add the mushrooms and onion and fry evenly.
3. Put the spaghetti, 150cc fresh cream, milk, and broth into the pan. Reduce until only half stock remains. Then add egg yolks and 150 grams of fresh cream. Simmer with moderate heat to thicken the stock. Season with salt, pepper (5:1) in the last stage. Finally, sprinkle with parsley.

焦糖布丁
Caramel Pudding

難易度 ★★　　　時間 100 分

材料 Ingredients

A	60g	糖	sugar
	10cc	水	water
B	350cc	全脂牛奶	milk
	2.5cc	香草精	vanilla extract
	150cc	無糖鮮奶油	cream
	125g	糖	sugar
	4 顆	蛋	eggs
	60cc	打發鮮奶油	whipped cream
	4 顆	紅櫻桃	cherry
	4 支	薄荷葉	pc mint leaf

作法 Method

1. 將材料A煮至成黃褐色即為焦糖；烤箱預熱165℃。
2. 布丁杯塗上薄薄一層奶油後，加入焦糖備用。
3. 將材料B中的糖與一半的牛奶煮至糖溶化後，續入剩餘的牛奶、鮮奶油及蛋攪拌均勻，過濾備用。
4. 將步驟3之布丁液倒入布丁杯中，蓋上鋁箔紙隔水放入烤箱以165℃烤40～50分鐘，至布丁液凝固後，取出放涼。
5. 置入冰箱中以5℃冷藏60分鐘，取出倒扣於盤上，擠上鮮奶油，以紅櫻桃及薄荷葉裝飾即可。

---

1. Make caramel with ingredients A ; preheat over to 165℃.
2. Butter of pudding mold then add caramel into it.
3. Cook the sugar (ingredients B) with 1/2 milk until the sugar is melted, then pour the rest milk, cream and egg and mix up equally then strain them to be custard liquid.
4. Pour custard liquid into prepared pudding mold, cover with foil, placed in hot water. Remove any bubbles. Bake at 165℃ for 45 min until done.
5. Keep it in the fridge until chill and put on a plate up-side-down, then garnish whipping cream, cheery and mint leaf.

兒童套餐

Set Menu for Mothers' Day / 母親節套餐

Bread Roll
餐包

Caesar Salad with Grilled Chicken Breast
凱撒沙拉與香烤雞胸

French Onion Soup
法式洋蔥湯

Black Pepper Beef and Cheese Frittata
黑胡椒牛肉乳酪烘蛋

Lemon Salmon Roll with Cream Mustard Sauce
檸檬鮭魚捲與奶油芥末醬

Butter Bread Pudding
英式麵包布丁

每年到了母親節或父親節，常會找家特別的餐館，來感謝父母親的辛勞，但往往因一位難求且人潮擁擠，不但餐飲及服務品質沒有隨著價位而提高，反而大打折扣，花了不少冤枉錢。在現今不景氣的年代，不妨在家 DIY 做套餐，更能表現孝心，且吃的健康，舒服，也更具意義。

這兩組菜單，可以說是西餐的經典，有爽口濃郁的凱撒沙拉配上法式洋蔥湯，主菜是地中海風味的檸檬鮭魚捲，甜點則選擇了甜而不膩的英式麵包布丁。到了夏天，適合作父親節餐點的有消暑的墨西哥海鮮沙拉，義大利最具代表性的蔬菜湯，主菜是口味稍重的黑胡椒菲力牛排，而提拉米蘇是最 Hito 的甜點。在餐點之中，搭配適合的葡萄酒，可使整套菜單更加完美出色。若主菜為紅肉類，可搭配紅葡萄酒佐餐，若為白肉或海鮮類，則可佐以白酒。

餐包
Bread Roll

難易度 ★★　　　時間 90 分

母親節套餐

材料 Ingredients

A 500g	高筋麵粉	bread flour
10g	新鮮酵母	fresh yeast
45g	白糖	sugar、2g 鹽 salt
1 顆	蛋	egg
150cc	水	water
80g	奶油	butter
適量	蛋液（蛋黃加水打散）	T.T. egg wash

作法 Method

1. 將材料A加入水攪拌均勻。
2. 揉5分鐘至麵糰有筋性。
3. 加入奶油，並揉5分鐘至表面光滑。
4. 蓋上布使其發酵30分鐘，分割成60g為1個的小麵糰，再醒30分鐘。
5. 刷上蛋液放入烤箱，以220℃的溫度烤25分鐘至呈金黃色即可。

. Mix up equally ingredients A with water.
2. Rub the dough for 5 minutes until it is soft.
3. Add butter and keep rubbing for another 5 minutes until the surface is smooth.
4. Then cover a cloth and let it ferment for 30 minutes, then cut it to be several small dough and 60g weight each and then rest for another 30 minutes.
5. Brush egg wash and put it in oven and bake at 220℃ for 25 minutes until golden.

凱撒沙拉與香烤雞胸
Caesar Salad with Grilled Chicken Breast

難易度 ★★　　　　　　時間 45 分

母親節套餐

材料 Ingredients

2 片	雞胸肉	pc chicken breast
1/2 顆	柳橙	orange
1/2 顆	檸檬	lemon
2 顆	蛋黃	egg yolks
300cc	橄欖油	olive oil
10cc	紅酒醋	red wine vinegar
10g	法式芥末子醬	French mustard seed
適量	鹽、胡椒	T.T. salt, pepper
2 顆	蒜頭	garlic
1 條	鯷魚	Anchovy
1 棵	蘿蔓生菜	Romaine
150g	培根碎	chopped bacon
	法國麵包（切6-8片）	baguette (cut to be 6 - 8 slices)
15g	帕瑪起司粉	Parmesan cheese powder
5cc	梅林烏醋	L. P. Sauce
2.5cc	辣椒水	Tabasco

香料奶油 Herb butter

30g	奶油	butter
3g	巴西利碎	chopped parsley
適量	鹽、胡椒	T.T salt, pepper

（奶油拌入巴西利碎打散，再加入鹽、胡椒調味。）
(Mix butter and chopped parsley equally then season with salt and pepper.)

作法 Method

1. 蒜頭切碎；培根炒香脆。
2. 法國麵包塗上香料奶油，烤上色備用。
3. 將雞胸肉洗淨，用柳橙汁、檸檬汁，鹽及胡椒醃約15分鐘，煎成金黃色至熟，保溫備用。
4. 將蛋黃打散，加入紅酒醋、法式芥末子醬，用打蛋器稍打發，再慢慢加入橄欖油使成濃稠狀，最後放入蒜頭、鯷魚、鹽及胡椒，續入梅林烏醋，以辣椒水調味即成凱撒沙拉醬。
5. 將蘿蔓生菜洗淨，拌入醬汁，再拌入培根碎及起司粉。
6. 將步驟3之雞胸肉切片，取3～4片裝盤，並附上香料奶油麵包即可。

---

1. Chop garlic; fry bacon to be crispy.
2. Spread herb butter on baguette and toast it.
3. Clean chicken breast then marinate it with orange juice, lemon juice, salt and pepper for about 15 minutes, then pan-fry it until the golden brown and well down, keep it warm.
4. Whip egg yolk, add red wine vinegar and mustard, and keep whipping and then add olive oil slowly to be slightly thickeded, and finally add garlic, anchovy, salt and pepper, then add L.P. sauce and season with Tabasco to be Caesar dressing.
5. Clean romaine then mix with dressing, and then mix up with chopped bacon and parmesan cheese.
6. Sliced chicken breast on the Caesar salad in a plate, with herb butter bread.

法式洋蔥湯
French Onion Soup

難易度 ★　　　　　　　　時間 75 分

母親節套餐

Memo

牛高湯亦可用雞高湯代替。
Could use chicken stock instead of beefstock.

材料 Ingredients

2 顆	洋蔥	onion
1 1/2 公升	牛高湯	liter beef stock
80g	奶油	butter
適量	麵粉、鹽、胡椒	T.T. flour, salt, pepper
40g	起司絲	grate cheese
4 片切片	法國麵包	pieces of baguett

香料包 Spice Bag

2 顆	黑胡椒粒	black pepper cone
1 顆	丁香	clove
1 顆	蒜頭	garlic
2.5g	迷迭香	rosemary
2.5g	百里香	thyme
1 片	月桂葉	bay leaf

作法 Method

1. 洋蔥切絲；蒜頭切碎；法國麵包塗上少許奶油備用。
2. 取一鍋子，待鍋熱後，放入奶油炒香洋蔥絲至洋蔥軟化成淺褐色，續加入 5g 麵粉，用小火炒至金褐色，再倒入高湯煮滾，轉小火加入香料包熬煮 45 分鐘，拿掉香料包，用鹽、胡椒調味。
3. 將湯盛入湯盤，加入 1～2 片法國麵包，撒上少許起司絲，放入烤箱以 200℃ 烤至起司融化呈金黃色即可。

---

1. Cut onion to be julienne; chop garlic; spread butter on baguette.
2. Sweat onion with butter until light brown colour, then add some flour, keep frying it until it turned to be golden colour, then put it into stock and spice bag, bring to simmer for 45 minutes, then take the spice bag away and season.
3. Put soup into soup bowl, add 1 - 2 pieces of baguette then add grate cheese and bake in oven at 200℃ until the cheese is melted.

黑胡椒牛肉乳酪烘蛋
Black Pepper Beef and Cheese Frittata

難易度 ★★　　　　　　　時間 45 分

材料 Ingredients

20g	橄欖油	olive oil
60g	黑胡椒牛肉（切丁）	black pepper beef (diced)
90g	洋菇（切丁）	mushrooms (diced)
60g	洋蔥（切碎）	onions (chopped)
180g	黃甜椒（切丁）	yellow bell pepper (diced)
180g	紅甜椒（切丁）	red bell pepper (diced)
6 顆	蛋（打散）	eggs (beaten)
20g	黑橄欖（切片）	black olives (sliced)
2g	巴西利（切碎）	parsley (chopped)
10g	帕馬森起司絲	parmesan cheese
60g	巧達乳酪絲	shredded cheddar cheese

作法 Method

1. 雞蛋打散過濾備用。
2. 熱鍋及熱油後將洋蔥及洋菇炒香，在加入黑胡椒牛肉、黃甜椒及紅甜椒拌炒均勻。
3. 再將蛋液倒入鍋中拌炒，至蛋半凝固在加入灑上乳酪、起司絲和巴西利。
4. 最後放入烤箱 180°C 烤約 10 分鐘至熟呈黃色即可。

---

1. Beat the eggs then filter.
2. Heat the pan with oil and saute the onions and mushrooms. Add black pepper beef, yellow bell pepper, and red bell pepper, then fry evenly.
3. Put the beaten eggs into the pan and turn into a semi-solid, then sprinkle cheese and cheese wire and parsley.
4. Bake in the oven at 180°C for about 10 minutes and cook until nicely on a yellow color in the end.

母親節套餐

檸檬鮭魚捲與奶油芥末醬
Lemon Salmon Roll with Cream Mustard Sauce

難易度 ★★★　　　　時間 60 分

母親節套餐

材料 Ingredients

	4 塊	鮭魚（每塊 120g）	pieces of Salmon (120g per each)
	45g	橄欖油	olive oil
	1 顆	紅蕃茄（去皮切丁）	tomato, peeled and dice
	1/2 顆	青椒切丁	green pepper, dice
	1/2 顆	黃椒切丁	yellow pepper, dice
	15g	紅蔥頭末	shallot, chopped
	10g	芥末子醬	mustard seeds
A	1/2 顆	柳橙皮末	chopped orange zest
	1/2 顆	檸檬皮末	chopped lemon zest
	少許	鹽、胡椒	T.T. salt, pepper
B	125cc	無糖鮮奶油	fresh cream
	60cc	白酒	white wine

作法 Method

1. 鮭魚去皮去骨，將魚肉切成長條狀，撒上材料 A 醃拌均勻備用。
2. 以保鮮膜將魚肉捲成糖果狀後，放入水中煮至表面凝固，待涼取出，以橄欖油煎至兩面呈金黃色（約 7 分熟）。
3. 奶油芥末子醬：以奶油 30g 炒香紅蔥頭碎，倒入白酒以小火加熱濃縮，加入鮮奶油煮開後，再放入芥末子醬、鹽、胡椒調味成醬汁。
4. 將鮭魚捲放入烤箱以 180℃ 烤約 8 分鐘至熟。
5. 以奶油芥末子醬打底，加上炒香的蕃茄、青椒及黃椒丁裝飾，將鮭魚捲上盤即可。

---

1. Take away the skin and bones of salmon, then cut it to be strips, spread ingredients A onto it and mix up equally to marinate.
2. Roll the salmon with wrap to be candy shape, and blanch until the surface is solidified, and then pan-fry it with olive oil until both sides until golden brown colour (approximately medium well).
3. Cream Mustard Sauce: Deep-fry chopped shallots with butter, pour white wine and reduce with moderate heat, then add fresh cream stirring until thicken and add mustard seeds, salt and pepper to be sauce.
4. Bake salmon roll at 180℃ in oven for about 8 minutes.
5. Pour cream mustard seeds sauce to be the base, garnish with fried tomato, green pepper and yellow pepper and then put the salmon roll.

英式麵包布丁
Butter Bread Pudding

難易度 ★★　　　　時間 100 分

材料 Ingredients

4 片	吐司（對切）		pc toast bread
120g	融化奶油		melted butter
30g	葡萄乾（泡蘭姆酒備用）		
	raisins (marinate with rum)		

A	120g	糖	sugar
	30cc	水	water
B	700cc	全脂牛奶	milk
	300cc	無糖鮮奶油	fresh cream
	160g	糖	sugar
	8 顆	蛋	eggs
	2g	肉桂粉	cinnamon powder

作法 Method

1. 將材料 A 煮成黃褐色焦糖備用；吐司沾融化的奶油備用；烤箱預溫 190℃。
2. 將材料 B 中的糖加入 250cc 牛奶煮至糖溶化後，續倒入剩餘的牛奶、鮮奶油及蛋攪拌均勻，過濾為布丁液備用。
3. 將焦糖倒在 9 吋模具上，鋪上奶油吐司及葡萄乾，再加入布丁液至 8 分滿，蓋上鋁箔紙。
4. 放入烤箱以 190℃，隔水烤約 70 分鐘。
5. 取出，待降溫倒扣上盤即可。

---

1. Cook ingredients A to be brown caramel; dip melted butter on toast; preheat oven to 190℃.
2. Add sugar (ingredients B) into 250cc of milk and cook it until the sugar is melted, then pour the rest of milk, fresh cream and egg and mix up equally, then strain it to be custard liquid.
3. Pour caramel in a 9 inches model, spread butter toast and raisins, then add custard liquid, then cover with foil.
4. Place in hot water. Bake at 190℃ until done about 70 min.
5. Put butter pudding in a plate after the temperature is warm.

Set Menu for Fathers' Day

父親節 套餐

Seafood Cocktail
海鮮沙拉盅

Minestrone with
義式蔬菜湯

Grilled Beef Tenderloin with Black Pepper Sauce
碳烤菲力牛排與黑胡椒醬

Chicken Cordon Bleu with Bake Tomato
藍帶雞排附香料蕃茄

Tiramisu
提拉米蘇

People nowadays dine at a unique restaurant to thank for their parents on Mothers' Day or Fathers' Day; however, on these days, restaurants are usually fully booked. Therefore the quality of the food and services goes down. People will spend much more money than what it is worth. In the slow economic situation nowadays, try to make set dinner your own. It could not only show your filial piety, but also make yourself healthier and more comfortable.

These 2 dishes are classic western menu; light Caesar salad with French onion soup, main course is Mediterranean lemon salmon roll, with English butter bread pudding. In summer, suitable dishes for fathers' day are Mexican seafood salad, Minestrone, the main course is sirloin steak with black pepper which is stronger flavour, and the most popular dessert is tiramisu. Having those dishes with good wines would make them taste better. You can drink red wines with red meat for main course, and drink white wines with white meat for main courses.

海鮮沙拉盅
Seafood Cocktail

難易度 ★　　　　　　時間 20 分

材料 Ingredients

125g	美生菜絲	lettuce, julienne
適量	巴西利	T.T. parsley
4 顆	黑橄欖，切片	black olives, sliced
1/2 個	檸檬，切角	lemon, cut to be wages
適量	鹽、胡椒	T.T. salt, pepper

A	45g	美乃滋	mayonnaise
	30g	蕃茄醬	ketchup
	10g	檸檬汁	lemon juice
	5g	酸豆末	chopped caper
	2.5cc	辣椒水	Tabasco

B	80g	鮪魚丁	diced tuna
	80g	花枝丁	diced squid
	8 隻	草蝦仁	shrimps
	4 顆	生干貝（對切）	scallops (cut to be two equal parts)
C	1/4 顆	洋蔥丁	diced onion
	1/4 顆	青椒丁	diced green pepper

作法 Method

1. 將材料 B 的海鮮及材料 C 分別燙煮備用。
2. 將材料 A 攪拌均勻，再加入材料 B 及 C，並用鹽、胡椒調味。
3. 取雞尾酒杯，將生菜絲墊底，再加上海鮮沙拉，最後加上檸檬角、黑橄欖片及巴西利裝飾即可。

---

1. Blanch seafood (ingredients B) and ingredients C separated.
2. Mix up equally ingredients A then add ingredients B and C then season with salt and pepper.
3. Spread lettuce julienne on the bottom of a cocktail glass then put seafood salad, and garnish with lemon wages, black olive slices and parsley.

義式蔬菜湯
Minestrone

難易度 ★   時間 45 分

材料 Ingredients

310g	去皮蕃茄罐	can peeled tomato
100g	培根	bacon
1/4 顆	洋蔥	onion
2 顆	蒜頭，切碎	garlic, chop
150cc	橄欖油	olive oil
60g	通心麵	macaroni
30g	蕃茄糊	tomato paste
50g	青豆仁	green pea, blanch
2 公升	雞高湯	litre chicken stock
30g	白豆或花豆（泡水 2 小時）	white bean / pinto bean sooking water 2 hours)
適量	鹽、胡椒、起司粉	T.T. salt, pepper, parmesan cheese powder

A
- 150g 青蒜 leek
- 150g 紅蘿蔔 carrot
- 150g 西芹 celery
- 150g 高麗菜 cabbage
- 150g 洋芋 potato

B 適量 迷迭香・奧勒岡・九層塔 月桂葉
T.T. rosemary, oregano, basil, bay leaf

作法 Method

1. 將材料 A 的蔬菜均切丁；洋蔥、去皮蕃茄及培根亦切丁。
2. 鍋熱加入橄欖油將洋蔥、培根及蒜頭炒香，放入材料 A 炒香，續入蕃茄糊炒開，再放入蕃茄丁、花豆，最後加入雞高湯煮開。
3. 加入香料材料 B 以小火煮 20 分鐘，放入通心麵煮 20 分鐘，以鹽、胡椒調味，用青豆仁裝飾。

1. Cut ingredients A to be dice; cut onion, tomatoes and bacon to be dice.
2. Saut?onion, bacon and garlic with olive oil in stockpots, then add ingredients A and saut?until light brown colour add tomato paste, diced tomato, beans and add chicken stock bring to boil.
3. Add herb ingredients B and macaroni, simmer for 20 minutes then season, and garlish with green peas.

碳烤菲力牛排與黑胡椒醬
Grilled Beef Tenderloin with Black Pepper Sauce

難易度 ★★★　　　時間 60 分

材料 / Ingredients

4 塊	菲力牛排（每塊 180g）	pieces of beef tenderloin (180g per each)
50g	奶油	butter
30g	粗粒黑胡椒	black pepper crushed
500cc	牛高湯	beef stock
180cc	紅酒	red wine
60g	紅蔥頭碎	chopped shallot
4 小朵	白花菜	small cauliflower
20g	中筋麵粉	flour
2 顆	蒜頭碎	chopped garlic
4 支	玉米筍	baby corn
各 1 顆	紅、黃、綠甜椒（切 4 片）	red pepper, yellow pepper, green pepper (each cut to be 4 pieces each one)
	白花菜、玉米筍和甜椒燙煮後沖冷水備用	blanch cauliflower, baby corn and sweet pepper
4 顆	洋芋	potatoes
1 支	蔥（切蔥花）	spring onion, chopped
30g	酸奶油	sour cream

麵糊 / ROUX

用小火將 20g 奶油及 20g 中筋麵粉炒約 2 分鐘使其散發出香氣，不須上色。

Sauteéd 20g butter and 20g flour with moderate heat for about 2 minutes until the smell comes out.

作法 / Method

1. 將粗粒黑胡椒放入烤箱以 180℃ 烤上色（約 1 分鐘）或乾炒 1 分鐘備用。
2. 將洋芋帶皮洗淨後用鋁箔紙包起，放入烤箱以 180℃ 烤約 1 小時，待熟取出備用。
3. 用奶油將紅蔥頭、蒜頭炒香後，放入黑胡椒拌炒 1 分鐘，續入紅酒濃縮至一半的量，再加入牛高湯，用鹽、胡椒調味即成黑胡椒醬（若不夠濃稠，可加麵糊煮 3 分鐘）。
4. 菲力牛排用鹽、胡椒調味，再用中大火煎至兩面呈金黃色，放入烤箱以 180℃ 烤 7 分鐘約 8 分熟。
5. 將配菜白花菜、玉米筍和三色甜椒用奶油炒香。
6. 烤洋芋用小刀劃十字，由內往外壓成洋芋花，再加上酸奶油、奶油及蔥花。
7. 最後將菲力牛排上盤，加上配菜及烤洋芋，再淋上黑胡椒醬汁即可。

---

1. Bake black pepper crushed in oven at 180℃ or pan-fry for 1 minute.
2. Clean potatoes with skin then pack with foil, bake at 180℃ for about 1 hour until down.
3. Saut?shallots and garlic with butter then put black pepper for 1 minute, and then pour red wine reduce then add beef stock and season with salt and pepper to be black pepper sauce (could add some roux and cook for 3 minutes if it is not thicken enough).
4. Season tenderloin with salt and pepper then pan-fry the both sides until medium, then put in oven at 180℃ for 7 minutes until medium well.
5. Saut?cauliflower, baby corn and bell peppers with butter and season to be side dish.
6. Put a gentle cross on baked potato then press it from inside to outside, and garnish with sour cream, butter and chopped spring onion.
7. Finally put tenderloin in a plate then add side dish vegetable and baked potato with black pepper sauce.

父親節套餐

藍帶雞排附香料蕃茄
Chicken Cordon Bleu with Bake Tomato

難易度 ★★★　　　　　　時間 60 分

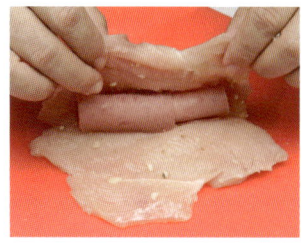

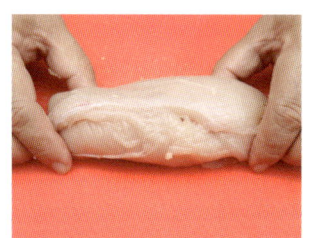

父親節套餐

材料 Ingredients

4 片 slices	雞胸肉	chicken breasts
4 片 slices	火腿片	ham
4 片 slices	起司片	cheese slices
2 個	蛋	eggs
150g	低筋麵粉	low-gluten flour
300g	麵包粉	bread crumbs
2 顆	蒜頭（切碎）	garlic (chopped)
60 cc	橄欖油	olive oil
2 顆	紅蕃茄	red tomatoes
3 g	奧立岡	oregano
50g	奶油	butter
50g	沙拉油	salad oil
適量	鹽及白胡椒	T.T. salt and white pepper

作法 Method

1. 烤箱預熱至180℃；雞胸肉切拍薄成0.5公分用鹽胡椒及蒜頭調味；起司片切條備用。
2. 紅蕃茄對切備用；100g麵包粉加入橄欖油及奧立岡拌攪均勻並用鹽胡椒調味，平均鋪在紅蕃茄上備用。
3. 火腿片包入起司條，再用雞胸肉片包捲起來，依序沾中筋麵粉、蛋液及麵包粉，平底鍋入奶油及沙拉油燒熱，將藍帶雞排煎至呈金黃色，最後與香料蕃茄一同入烤箱烤180°C約12分鐘即可。

1. Preheat oven to 180 ℃. Split chicken breasts into 0.5 cm thin. Season with salt, pepper and garlic. Cut cheese slices into strips.
2. Cut all the red tomatoes in half, and set aside. 100 grams of bread flour add with olive oil and oregano, then stir evenly, season with salt and pepper, then spread over red tomatoes evenly, and set aside.
3. Cheese cut into strips wrapped in ham slices, and rolled up with chicken breast slices. Sequentially dip low-gluten flour, egg and bread crumbs. Heat the frying pan with butter and salad oil, and fry the Chicken Cordon Bleu to golden brown, then put into the oven with spice-tomatoes at 180°C for about 12 minutes in the end.

提拉米蘇
Tiramisu

難易度 ★★　　　　　　　時間 60 分

父親節套餐

材料 Ingredients

1 顆	蛋黃	egg yolk
25g	糖粉	icing sugar
1 片	吉利丁	piece of gelatin
30cc	蜂蜜	honey
30cc	白蘭地	brandy
125g	馬士卡彭起司（打散）	Mascarpone Cheese
125g	無糖鮮奶油打發	whipped fresh cream
30cc	咖啡酒	kahlua
1 個	5 吋海綿蛋糕	5 inches sponge cake
15g	可可粉	coco powder
4 顆	草莓	strawberries

作法 Method

1. 將蛋黃及糖粉隔熱水打散，至糖粉溶化。
2. 吉利丁泡冰水待軟，過濾，加入蛋黃中打散。
   加入蜂蜜、白蘭地拌均勻。
3. 續加入馬士卡彭起司及打發的無糖鮮奶油，攪拌均勻。
4. 在模型杯上先放一層海綿蛋糕為底，刷上少許咖啡酒，倒入提拉米蘇高度約 1.5 公分，再鋪上另一層海綿蛋糕，並刷上咖啡酒，最後再倒入剩下的提拉米蘇至容器約 9 分滿。
5. 將其放入冰箱以 3℃ 冷藏約 1 小時，撒上可可粉再飾以草莓即可。

1. Whipped egg yolk and icing sugar over hot water until the icing sugar is melted keep warm.
2. Soaked gelatin in iced water to be soft, and then strain it and add to egg yolk until melted. Mix up equally with honey and brandy.
3. Then add Mascarpone cheese and whipped cream and mix up equally.
4. Put one layer of sponge cake to be the base in a model cup, brush coffee liqueur then pour into tiramisu for about 1.5cm, and then put another layer of sponge cake and brush coffee liqueur, at last pour into the rest of tiramisu until it is about 9/10 of the container.
5. Fridge to chill at 3℃ for about 1 hour, then spread coco powder and garnish with strawberry.

Christmas Fruit Set Menu
聖誕節水果套餐

Spiced Pears in Red Wine Sauce
肉桂紅酒梨

Toasted Corn and Sweet Squash Soup
奶油玉米甜瓜湯

Gluehwein
德國聖誕紅酒

Lomi Lomi Salmon
橙香夏威夷鮭魚沙拉

Roasted Turkey Breast with Fruit Stuffing
香烤果餡火雞捲

Snow Ball
酪梨香草雪球

白色聖誕！！多麼令人期待的日子呀！在歐美視為全家團聚的節日，而在台灣多數人為了迎接這天的到來，訂餐廳、開 Party 大費周章。其實，你也可以與親朋好友在聖誕樹旁享受片刻的溫馨，在家 DIY 烹調出連餐廳也享受不到的特別美味。時至今日，台灣的農業技術已執世界牛耳的地位，好吃的水果垂手可得，即使現代人飲食講求天然環保，但在歡樂的節慶裡，還是會忽略了以健康為最終的訴求，所以，此套菜單特別以水果入菜，搭配鮭魚，南瓜，火雞，冰淇淋等食材，讓你在享受美味的同時，也為健康把關。

White Christmas, what an expecting day it is!! It is a holiday for the whole family get together in westren world, in Taiwan, most people make reservations in restaurants to celebrate. Actually, you can enjoy the moment with your family, friends behind the Christmas tree, the wonderful experiences that you'll never have in restaurants. Nowadays, Taiwanese agricultural skills reaches the top stage in the world, although people eat daintily especially for the environment protection, we still usually forget having good food to keep healthy in the happy holidays. Therefore, we create this menu which contains fruit with salmon, pumpkin, turkey and ice cream. Those dishes can not only make good taste but also keep you healthy.

肉桂紅酒梨
Spiced Pears in Red Wine Sauce

難易度 ★　　　　時間 105 分

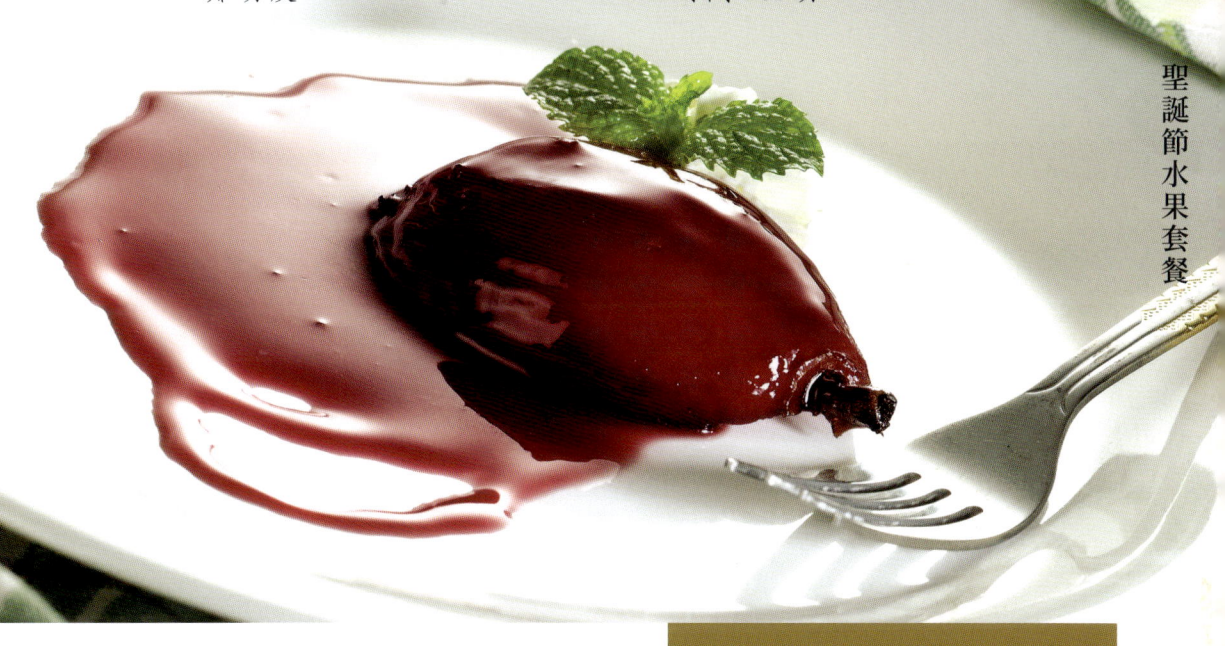

聖誕節水果套餐

材料 Ingredients

500cc	紅酒	red wine
500g	糖	sugar
1 支	肉桂棒	cinnamon
2 顆	丁香	cloves
1/2 顆	檸檬皮	lemon zest
1/2 顆	柳橙皮	orange zest
4 顆	鴨梨（去皮，泡鹽水）	pears (peeled, soaked salt water)
15g	玉米粉（加入 190cc 水調勻，芶芡用）	corn starch
適量	無糖鮮奶油打發	T.T. whipped cream
4 片	薄荷葉	mint leaves

※ 若鴨梨太大顆，則可對切使用。鴨梨以進口居多，有紅皮及綠皮之分，又以紅皮為佳，亦可使用本地的水梨代替。

作法 Method

1. 將紅酒、糖、丁香、肉桂棒、檸檬皮及柳橙皮放入小湯鍋中，煮至沸騰後轉小火加蓋再慢煮 15 分鐘。
2. 放入鴨梨，使紅酒淹蓋過鴨梨，以慢火續煮 30 分鐘。
   移火，待鴨梨冷卻為室溫約 1 小時後取出鴨梨。
3. 將紅酒過濾，濃縮 8～10 分鐘成約 250cc 的量，續入玉米粉加水芶芡，即為紅酒醬汁。
4. 將紅酒梨上盤淋上醬汁，擠少許鮮奶油，再用薄荷葉裝飾即可。

1. Cook red wine, sugar, cloves, cinnamon stick, lemon and orange zest until they are boiled and simmer for 15 minutes.
2. Put pear into the pot and be covered by red wine, simmer for 30 minutes.
   Turn off the heat, after pear gets cooled and keep room temperatures.
3. Strain red wine and reduce for 8～10 minutes to be 250cc sauce left, then add corn starch with water and adjust texture to be red wine sauce.
4. Put red wine pear and sauce in a plate, garnish whipping cream and mint leaves.

材料 Ingredients

200g	地瓜	sweet potato
400g	南瓜	pumpkin
1 支	甜玉米	sweet corn
125cc	鮮奶油	fresh cream
適量	鹽、胡椒	T.T. salt, pepper
30g	奶油	butter

A	100g	洋蔥	onion
	50g	紅蘿蔔	carrot
	10g	薑片	ginger slices
B	1 公升	雞高湯	litre chicken stock
	250cc	牛奶	milk
	125g	玉米醬	corn sauce

作法 Method

1. 將地瓜及南瓜去皮、去籽、切丁；洋蔥及紅蘿蔔切丁；甜玉米入烤箱烤上色，取下玉米粒，玉米梗備用，預留 1/2 杯玉米粒裝飾用。
2. 用奶油將材料 A 炒香，續入地瓜、南瓜炒至表皮上色，再加入玉米梗，最後入材料 B 以大火煮開，轉小火煮至南瓜鬆軟（約 20 分鐘）。
3. 取出玉米梗，以果汁機打成泥狀，過濾，加入鮮奶油煮至沸騰，再用鹽及胡椒調味，最後用玉米粒裝飾即可。

---

1. Peeled and seeded of sweet potato and pumpkin, and dice them; dice onion and carrot; bake sweet corn until light brown colour and take away corn, 1/2 cup for garnish and keep corn stalk.
2. Saut?the ingredients A with butter then put sweet potato and pumpkin, until light brown colour, then add corn stalk and ingredients B, and simmer until the pumpkin becomes soft (about 20 minutes).
3. Take corn stalk out and pure?soup by blender, strain it, add fresh cream bring to boil and season and garnish with sweet corn.

奶油玉米甜瓜湯
Toasted Corn and Sweet Squash Soup

難易度 ★　　　時間 45 分

聖誕節水果套餐

德國聖誕紅酒
Gluehwein

難易度 ★　　　　　時間 45 分

材料 Ingredients

1 瓶	紅葡萄酒 (750cc)	
1 個	香吉士或柳丁	sunkist or oranges
5 顆	丁香	cloves
1	肉桂棒	cinnamon stick
1 顆	八角 (小)	anise (small)
2 片	薑片	ginger slices
60g	白糖	white sugar
150cc	柳橙汁	orange juice

作法 Method

1. 香吉士釘上丁香並將所以材料小火悶煮 30 分鐘即可。

1. Push the cloves into the oranges, then simmer all the ingredients for 30 minutes and serve.

59

橙香夏威夷鮭魚沙拉
Lomi Lomi Salmon

難易度 ★★　　　　　　　　時間 110 分

Ingredients 材料

500g	鮮鮭魚	fresh salmon
300g	洋蔥（約 1 顆）	onion
300g	紅蕃茄（約 1 顆）	tomato
1/3 顆	美生菜	lettuce
10g	巴西利	parsley
150g	紅甜椒	red bell pepper
150g	黃甜椒	yellow bell pepper

A	1 1/2 顆	檸檬壓汁	lemon juice
	1 1/2 顆	柳橙汁	orange juice
	10g	鹽	salt
	10g	糖	sugar
	4 滴	辣椒水	drops of Tabasco
	150cc	鳳梨汁	pineapple juice

作法 Method

1. 洋蔥切絲；紅蕃茄去籽切絲；紅、黃甜椒洗淨均切絲；巴西利切碎備用。
2. 鮮鮭魚切長條片狀（約 0.5×1×9 公分），放入洋蔥絲及材料 A，醃 90 分鐘。
3. 將醃漬鮭魚拌入紅、黃甜椒絲及紅蕃茄絲醃泡約 15 分鐘至蔬菜入味。
4. 以美生菜放在盤上墊底，加上醃漬鮭魚，再放些巴西利碎及鳳梨葉裝飾即可。

---

1. Seeded of tomato and cut tomato, onion and bell pepper it to be julienne; chop parsley.
2. Cut fresh salmon to be strips (about 0.5×1×9 cm), add onion julienne and ingredients A and marinate for 90 minutes.
3. Mix marinated salmon with tomato and bell pepper and tomato julienne and keep marinating for 15 minutes.
4. Put lettuce on the bottom of a plate, then put marinated salmon and garnish with chopped parsley and pineapple leaves.

※ 選購鮭魚時，一定要新鮮，亦可用鮪魚或旗魚代替。
Choosing fresh salmon to make this dish is necessary. You can choose tuna or other fish to instead of salmon.

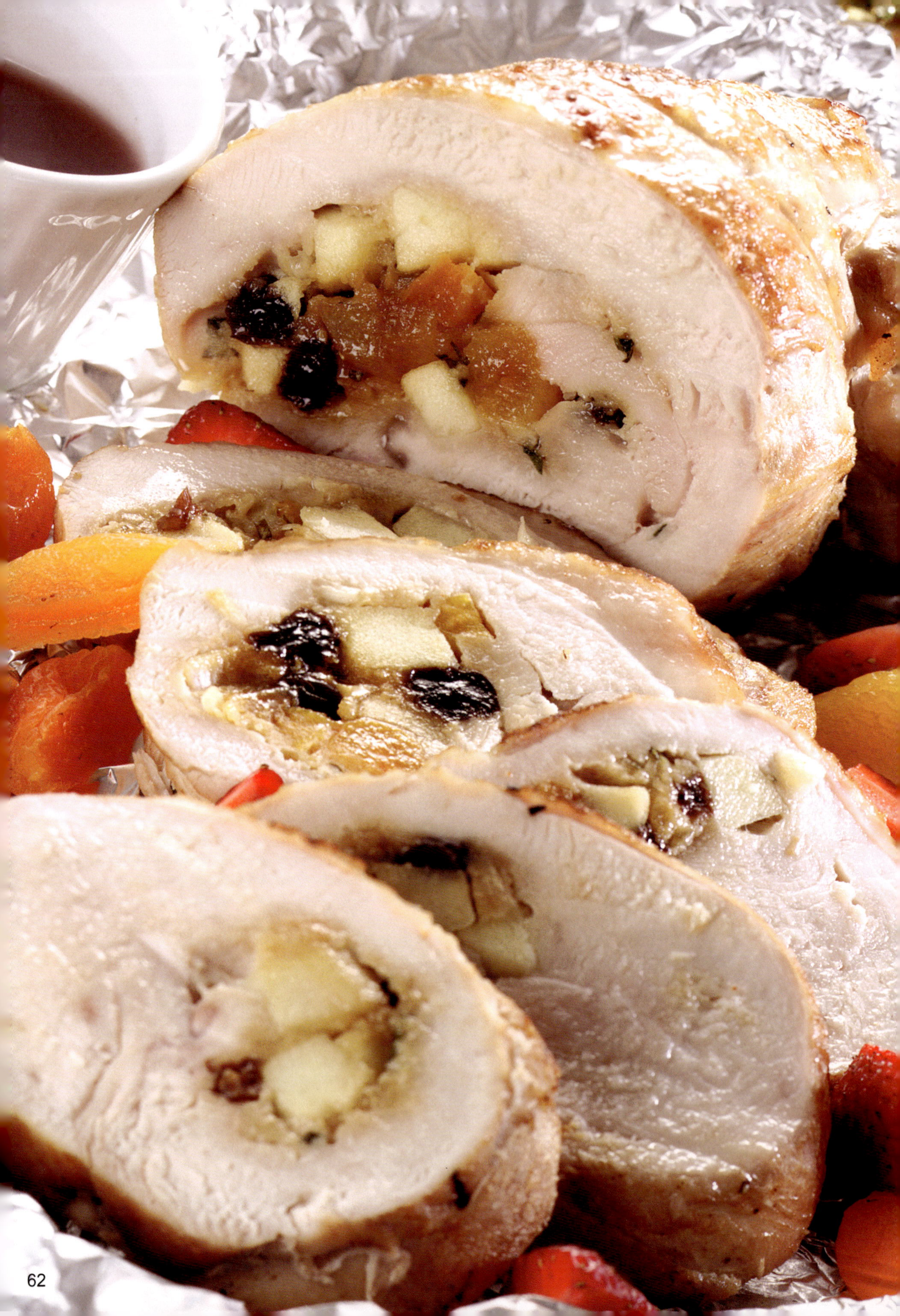

香烤果餡火雞捲
Roasted Turkey Breast with Fruit Stuffing

難易度 ★ ★ ★　　　　　時間 80 分

材料 Ingredients

1kg	火雞胸	turkey breast
30g	蔬菜油	vegetable oil
15g	胡椒鹽	pepper, salt
2 條	綿繩	cotton rope
	水果餡	Fruit stuffing
30g	奶油	butter
150g	洋蔥碎	chopped onion
1 顆	青蘋果	green apple
10g	紅糖	brown sugar

果香醬汁 Fruit sauce

A	60g	杏桃乾	dried apricot
	30g	蔓越莓乾或葡萄乾	dried cranberry or raisin
	30cc	蘋果汁	apple juice
	30cc	雞高湯	chicken stock
	30cc	紅酒醋	red wine vinegar
	5g	鼠尾草	sage
	2 顆	丁香碎	chopped cloves
B	125cc	蘋果汁	apple juice
	80cc	雞高湯	chicken stock
	85cc	紅酒醋	red wine vinegar
	30cc	白酒	white wine
	30g	洋蔥切碎	chopped onion

聖誕節水果套餐

作法 Method

1. 將火雞胸肉展開拍打成長方形狀，撒上鹽、胡椒調味備用。
2. 青蘋果去皮、切丁；杏桃乾切丁；鼠尾草、洋蔥切碎；烤箱預熱至 180℃ 備用。
3. 水果餡：用奶油將洋蔥炒香後，加入蘋果及紅糖炒至蘋果表皮上色，再加入材料 A 以小火煮至汁液收乾，熄火降溫備用。
4. 將水果餡放在展開的火雞胸肉上，捲成筒狀，並用綿繩固定，用少許油，將火雞捲以中火煎上色，再放入烤箱烤至熟（約 20～25 分鐘）。
5. 果香醬汁：將烤剩餘的肉汁及油，加入材料 B 煮至滾，轉小火並濃縮至約 250cc 的量，加入鹽、胡椒調味。
6. 除去火雞捲的綿繩，切成約 1.5 公分的厚片，再淋上醬汁即可。

1. Flatten turkey breast with a meat mallet, and then season.
2. Peeler green apple then dice it; dice dried apricot; chop sage and onion; Preheat oven to 180℃.
3. Fruit stuffing : saut?onion with butter then add apple and brown sugar, until golden brown colour, then add ingredients A and reduce until the juice is absorbed then turn off the heat.
4. Put fruit stuffing on the turkey breast, then roll it and tie with cotton rope, pan-fry the turkey roll until golden brown colour, then put in oven until done (about 20 ~ 25 minutes).
5. Fruit sauce: cook the rest of gravy, oil and ingredients B bring to simmer and reduce then season.
6. Take away the rope then cut turkey roll to be slices which are about 1.5cm thick each, then pour sauce on them.

酪梨香草雪球
Snow Ball

難易度 ★　　　　　時間 30 分

材料
Ingredients

0.6litre	香草冰淇淋	vanilla ice cream
1/2 個	新鮮酪梨切丁	avocado, diced
60cc	檸檬汁	lemon juice
適量	荳蔻粉	T.T. nutmeg powder
310cc	特調檸檬汁（15g果糖及檸檬汁加 250cc 冰水調均勻）

lemon juice cooler(mix up 15g of syrup and lemon juice with 250cc iced water)

作法
Method

1. 將 1 杯香草冰淇淋及酪梨、檸檬汁，用果汁機打散。
2. 將剩下的香草冰淇淋裝入杯中，淋入酪梨汁及特調檸檬汁，再撒上荳蔻粉，以檸檬皮絲及柳橙裝飾即可。

1. Blend 1 cup of vanilla ice cream, avocado and lemon juice in a blender.
2. Put the rest of vanilla ice cream into a cup, add avocado juice and lemon juice cooler, and garnish with nutmeg and zest and orange wedge.

※ 酪梨要選購軟的，若太生，可放入米缸中 1～2 天。

Chinese New Year Set Menu 1 中國新年 套餐一

Creamed Oyster and Shrimp in Shell
焗奶油明蝦生蠔

Pork Ribs with B. B. Q. Sauce
豬肋排附美式烤肉醬

Chocolate Fondue with Fresh Fruit
水果巧克力火鍋

Mushroom and Lotus Seeds Cappuccino Soup
蘑菇蓮子卡布奇諾湯

利用年貨的食材，配合西式的烹調法，在家也能做出有年味的西式套餐。西式的年菜較感覺不出濃厚的年節氣氛，所以，在此特別設計了兩道圍爐菜，一道是含有濃郁蕃茄的義式火鍋，搭配鮮美的海鮮。

另一道則選擇以飯後的巧克力火鍋配上新鮮水果，既不失團圓圍爐的氣氛，更增添幾分巧思。在食材方面也可有所變化，中式烤鴨可做出爽口不油膩的華爾道夫沙拉，蓮子與洋菇也成了濃郁的卡布奇諾湯，加上金桔做成的奶酪，是最應景不過的甜點了！

You can make Chinese New Year Dinner with west style by Chinese ingredients and western cooking way. It is hard to feel the New Year from Western New Year dishes, therefore, we set up two special dishes for it, one is Italian fondue with tomato and fresh seafood, the other one is chocolate fondue with fresh fruit which is a dessert. There are some changes for ingredients: Chinese BBQ duck Waldorf Salad、rich cappuccino soup with mushroom and lotus seeds、Panna Cotta made by Kumquat.

焗奶油明蝦生蠔
Creamed Oyster and Shrimp in Shell

難易度　★　　　　　　時間 45 分

材料 Ingredients			
90g	奶油	butter
8 隻	草蝦（40g／隻）	grass shrimps (40g each)
8 個	生蠔	oysters
250cc	牛奶	milk
45g	中筋麵粉	flour
20g	紅蔥頭碎	chopped shallot
1 顆	蛋黃	egg yolk
適量	鹽、白胡椒	T.T. salt, white pepper
80g	麵包粉	bread crumbs
120g	起司絲	grate cheese
1/2 顆	檸檬角	lemon wages

簡易高湯 Court Bouillon		
125g	洋蔥塊	onion cubes
100g	西芹塊	celery cubes
1 片	月桂葉	bay leaf
90cc	白酒	white wine
1L	水	water

作法 Method

1. 烤箱預熱 210℃。
2. 將草蝦去頭及殼，取出沙筋，預留蝦頭及 1/3 的蝦尾，剩餘蝦肉切小丁備用，草蝦頭洗淨備用。
3. 生蠔取肉，將殼洗淨備用。
4. 簡易高湯：1 公升水加入白酒及調味蔬菜，煮至沸騰，轉小火煮約 20 分鐘濃縮，加入生蠔燙煮約 30 秒至 5 分熟，備用。
5. 將煮液濃縮過濾約 250cc。
6. 熱鍋，放入 15g 奶油，將紅蔥頭炒香，再加入蝦仁丁，炒至上色，淋入少許白酒拌炒，取出蝦肉。
7. 用同一炒鍋加 30g 奶油及 30g 麵粉炒成麵糊，再加入牛奶及煮液攪拌至無顆粒，濃縮成濃稠狀，續加入蛋黃及炒香的蝦仁丁，即為蝦仁奶油焗醬。
8. 將生蠔放入殼中，再放入 1/3 隻草蝦及頭裝飾，淋上蝦仁奶油焗醬後，鋪上起司絲及麵包粉。
9. 送入烤箱焗烤 10～15 分鐘至呈金黃色即可。亦可附上檸檬角。

---

1. Preheat oven to be 210℃.
2. Clean and peeled grass shrimp, keep shrimp heads and 1/3 of shrimp tails and dice the rest of shrimp meat.
3. Take the oyster out of the shell and clean the shells.
4. Make court bouillon with white wine and mirepoix, blanch oyster about 30 sec, and strain.
5. Reduce court bouillon until good flavor for stock.
6. Sautéed shallot with butter and then add shrimp until light brown colour, add white wine to fry with, then take out the shrimps.
7. Make roux with butter and flour, add milk stirring until thicken and then add egg yolk and shrimp to be creamy shrimp sauce.
8. Place oysters into shells then put 1/3 of shrimp tails and head to be garnish, add creamy shrimp sauce and spread cheese and bread crumbs.
9. Bake it in oven for 10 ~ 15 minutes until the golden brown colour. Serve with lemon wage.

中國新年套餐一

豬肋排附美式烤肉醬
Pork Ribs with B B Q Sauce

難易度 ★★　　　　時間 105 分

美式烤肉醬 材料 Ingredients for B.B.Q. sauce

500g	蕃茄醬	ketchup
90cc	醋	vinegar
45cc	濃縮柳橙汁	concentrated orange juice
90g	糖	sugar
5g	白胡椒粉	white pepper
5g	洋蔥粉	onion powder
5g	大蒜粉	garlic powder
適量	辣椒水	T.T. Tabasco
5cc	燻煙汁	smoked liquid

美式烤肉醬 作法 Method for B.B.Q. sauce

1. 將白醋跟糖混合煮滾，加入蕃茄醬及濃縮柳橙汁攪拌均勻。
2. 加入白胡椒粉、蒜頭粉、洋蔥粉拌均勻，用慢火煮20分鐘。
3. 加入適量的辣椒水及煙燻汁調味即可。

1. Mix whiter vinegar, sugar, ketchup and concentrated orange juice bring to boil.
2. Add white pepper, garlic powder and onion powder and simmer for 20 minutes.
3. Add Tabasco and smoked liquid to taste.

豬肋排 材料 Ingredients

400g	豬肋排 4 塊	Pork rib 4 pc
30cc	煙燻汁	smoked liquid

豬肋排 作法 Method

1. 烤箱預溫 200℃。
2. 將 30cc 煙燻汁以 300cc 水稀釋成醃漬液，將豬肋排浸泡於煙漬液中約 20 秒取出。
3. 在烤盤中加入熱水，放上浸泡過的肋排，蓋上鋁箔紙，放入烤箱約 70 分鐘燜烤至熟透。
4. 最後再與烤肉醬一起燜煮 25 分鐘至入味即可。

1. Preheat oven to 200℃.
2. Mix 30cc smoked liquid with 300cc water to be marinate liquid, then soaked pork rib into the marinate liquid for about 20 seconds and take out.
3. Add hot water in a baking tray and put pork rib then cover with foil and put in oven for about 70 minutes until tender.
4. Brais pork rib with BBQ sauce for another 25 minutes until done.

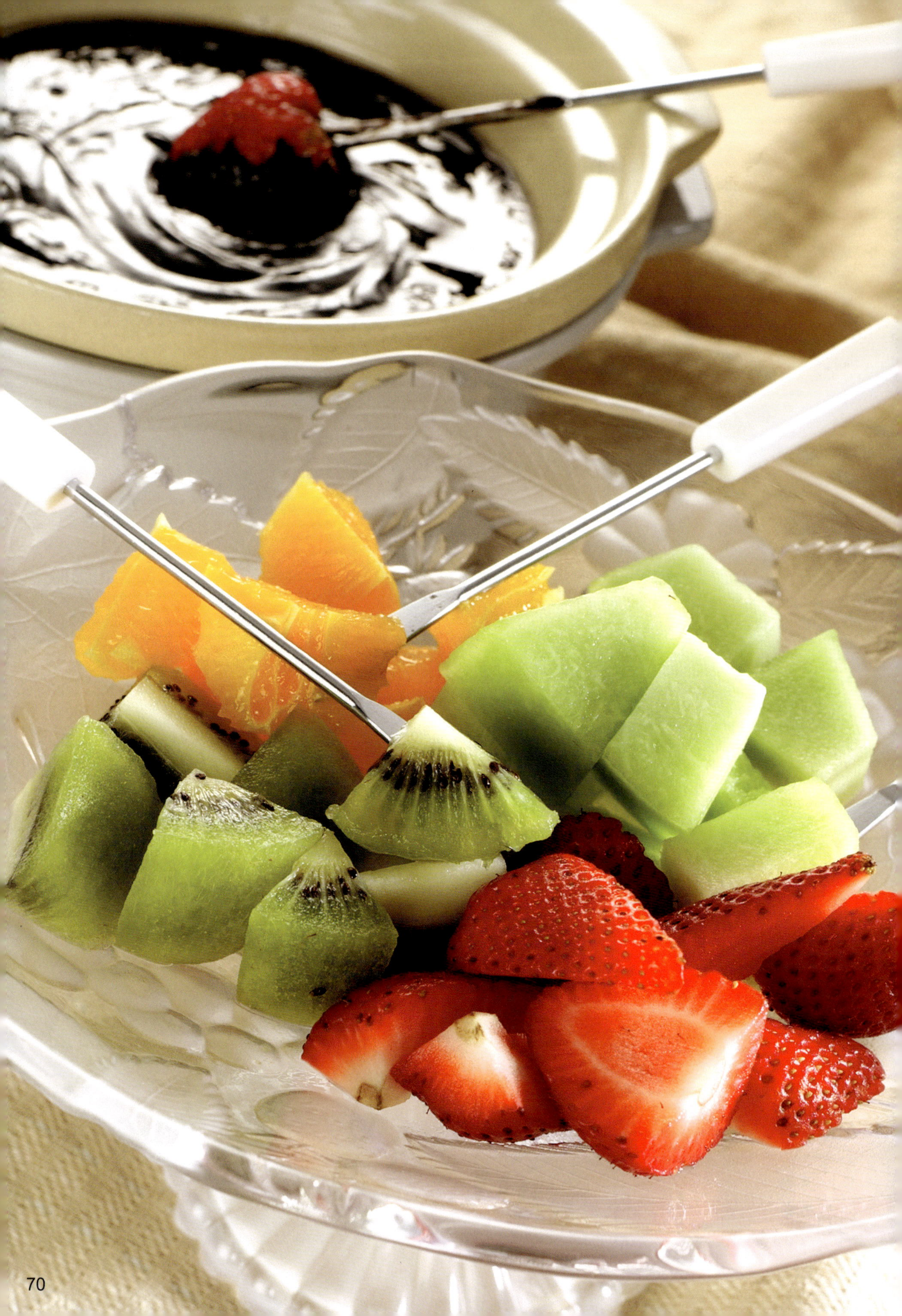

水果巧克力火鍋
Chocolate Fondue with Fresh Fruit

難易度 ★　　　　　　　　　　　　時間 30 分

中國新年套餐一

材料 Ingredients

360g	苦甜巧克力	dark chocolate
250cc	無糖鮮奶油	fresh cream
30cc	櫻桃白蘭地	cherry brandy
15g	咖啡粉	instant coffee

水果 Fruit

4 顆	奇異果	kiwifruits
8 顆	草莓	strawberries
1/2 顆	蜜世界	melon
4 顆	柳橙（去皮）	oranges (peeled)

作法 Method

1. 咖啡粉加 50cc 熱水調成濃縮咖啡。
2. 將巧克力隔水加熱至融化，再加入鮮奶油、櫻桃白蘭地及濃縮咖啡拌均勻。
3. 將切塊的水果排列整齊上盤。
4. 將巧克力醬放入 Fondue 鍋中，隔水以小火保溫，再附上綜合水果盤即可。

---

1. Make strong coffee with 50cc hot water and instant coffee.
2. Melt chocolate over hot water, then add cream, cherry brandy and coffee and mix well.
3. Set up all fruit cubes in a plate.
4. Place melted chocolate into the fondue pot, keep it warm with moderate heat, then set up the assorted fruit platter on the side.

※ 保溫巧克力的水溫勿過高，否則容易油水分離。
　 The temperature chocolate should not be too high otherwise it would make the chocolate separate.

※ 水果種類可依個人喜好而改變，以新鮮為佳。
　 You can choose all kinds of season fruit.

蘑菇蓮子卡布奇諾湯
Mushroom and Lotus Seeds Cappuccino Soup

難易度 ★　　　　　時間 45 分

材料 Ingredients

350g	新鮮洋菇切片	mushroom, sliced
1/2 顆	洋蔥碎	chopped onion
100g	生蓮子	lotus seeds
60cc	白酒	white wine
200g	洋芋（去皮切丁）	potato, peeled and diced
60g	奶油	butter
1 1/2 公升	雞高湯	litre chicken stock
125cc	無糖鮮奶油	fresh cream
250cc	牛奶	milk
適量	鹽、白胡椒、百里香	T.T. salt, pepper, thyme

作法 Method

1. 用奶油炒香洋蔥，再加入洋菇片炒至軟化，預留 1/4 裝飾用。
2. 加入百里香、生蓮子及白酒濃縮 1 分鐘。
3. 加入洋芋及雞高湯，煮至沸騰後轉小火，再煮約 20 分鐘至洋芋及蓮子熟透，取出降溫。
4. 放入果汁機中打成濃稠無顆粒狀，加入鮮奶油及預留的蘑菇片，攪拌均勻，用鹽、胡椒調味。
5. 將牛奶置於鍋中用慢火煮，再用打蛋器將牛奶打成牛奶泡沫。
6. 湯杯中倒入約 3/4 杯的蘑菇湯，再加上牛奶泡沫及少許蘑菇片裝飾即可。

1. Sweat onion with butter then add mushroom and until soft, keep the 1/4 of it for garnish.
2. Add thyme, rare lotus seeds and white wine than reduce for 1 minute.
3. Add potato and chicken stock, bring to simmer for about 20 minutes until done.
4. Pure?soup and add fresh cream and the rest of mushroom slices then season to taste.
5. Whip milk in a pot and cook to be milk form.
6. Pour 3/4 cup of soup in a cup, and then add milk form and few mushroom slices for garnish.

Chinese New Year Set Menu 2
中國新年 套餐二

Pear Waldorf Salad with Roasted Duck
烤鴨香梨華爾道夫沙拉

Pesto Sauce
青醬

Italian Tomato Seafood Fondue
義式蕃茄海鮮鍋

Slices Boneless Beef Short Ribs with Mushroom Brandy Sauce
無骨牛小排附蘑菇白蘭地醬

Kumquat Panna Cotta
金桔奶酪

主菜更是豐富，有大紅喜氣的美式烤豬肋排、香醇的白蘭地牛小排及美味的焗明蝦生蠔。

希望這兩套菜單的搭配組合，讓你在準備年菜宴客時，也能換換新口味，更添喜氣。

The main course is more fantastic: American BBQ pork rib、beef short rib with cognac and gratened shrimp with oysters. Hopefully these two combinations would help you to make wonderful New Year dinner.

烤鴨香梨華爾道夫沙拉
Pear Waldorf Salad with Roasted Duck

難易度 ★　　　　　時間 40 分

材料 Ingredients

1/2 隻	烤鴨（去骨切片）	roasted duck (boneless and sliced)
1 顆	紅鴨梨	red pear
1 顆	蘋果	apple
2 支	西芹	stalk of celeries
適量	鹽、胡椒	T.T. salt, pepper
150g	豌豆苗	pea shoot
3g	巴西利	parsly
40g	烤過核桃	roasted walnut
40g	蔓越莓乾	dried cranberry
90g	美乃滋	mayonnaise
50g	杏桃乾切丁	dried apricot, diced
75g	酸奶油	sour cream
20cc	檸檬汁	lemon juice

作法 Method

1. 將西芹切斜片備用；蘋果及紅鴨梨去皮切片泡鹽水備用。
2. 調味美乃滋：美乃滋、酸奶油及檸檬汁一起攪拌均勻。
3. 取 1/2 鴨肉加入紅鴨梨、蘋果、西芹、蔓越莓乾及巴西利碎，再加入調味美乃滋攪拌均勻，以適量的鹽、胡椒調味。
4. 取豌豆苗墊底，擺上沙拉，用 1/2 的鴨肉片圍邊，再加少許的蔓越莓乾、杏桃乾及核桃裝飾即可。

---

1. Sliced celeries and blanch, peeled of apple and red pear, then cut to be slices and soaked into salt water.
2. Mayonnaise sauce: mix up equally mayonnaise, sour cream and lemon juice.
3. Take half duck meat and add red pear, apple, celery, dried cranberry and chopped parsley and then add mayonnaise sauce to mix up equally, then season to taste.
4. Put peas sprout to be the base and place salad on it, then set the edge with the another half duck meat and garnish with cranberry, apricot and walnut.

青醬
Pesto Sauce

難易度 ☆　　　　　時間 15 分

中國新年套餐二

材料 Ingredients

50g	九層塔	basil leaves
20g	蒜頭	garlic
15g	松子	pinenut
250cc	橄欖油	olive oil
適量	鹽、胡椒	T.T. salt, pepper
45g	帕瑪森起司粉	Parmesan cheese powder

作法 Method

1. 將所有材料放入果汁機中打成泥狀，再以鹽、胡椒調味即可。

1. Pure?all ingredients by blender then season with salt and pepper.

※ 做好的青醬可以冷藏保存 48 小時或冷凍保存。
　　Pesto sauce could be chilled for 48 hours or freeze inside of fridge.
※ 亦可拌義大利麵、沙拉或海鮮，為義大利菜不可或缺的醬汁。
　　It is a popular sauce in Italian cuisine which is able to mix with pasta, salad or seafood dish.

75

義式蕃茄海鮮鍋
Italian Tomato Seafood Fondue

難易度 ★★　　　　　　　　　時間 50 分

材料 Ingredients

1 隻	鱸魚	sea bass
6 隻	草蝦	grass shrimps
6 顆	大蛤	clams
6 顆	生干貝	scallops
1/2 棵	青花菜	broccoli
6 支	玉米筍	baby corn shoots
6 顆	洋菇	mushrooms
6 顆	小蕃茄	cherry tomato
60cc	橄欖油	olive oil
15g	蒜頭切碎	chopped garlic
150g	洋蔥碎	chopped onion
125cc	白酒	white wine
500cc	去皮蕃茄罐	peeled canned tomato
1 支	紅辣椒，去籽切碎	red chili, seeded and chop
1 支	巴西利 parsley	

中國新年套餐二

作法 Method

1. 鱸魚去骨去皮、切片，留魚頭及魚骨備用。
2. 草蝦去頭去殼備用；大蛤吐沙備用；生干貝洗淨備用。
3. 青花菜切小花，玉米筍、洋菇及蕃茄對切。
4. 以橄欖油炒香蒜頭及洋蔥，再加入蝦頭，炒上色後加白酒濃縮 1 分鐘，再加入去皮蕃茄罐及紅辣椒。
5. 加入 2 公升的水及魚頭煮至沸騰，轉小火，加入巴西利梗煮 30 分鐘倒出、過濾，再以鹽及胡椒調味即為蕃茄海鮮鍋湯底。
6. 排上海鮮盤、蔬菜盤及蕃茄海鮮鍋，另附青醬為沾醬。

. Cut fillet of sea bass then slice it, keep the head and bones for fish stock.
2. Take away the heads and shells of shrimps; clean scallops and clams.
3. Clean and dice broccoli, baby corn shoots, mushrooms and tomatoes.
4. Saut?garlic and onion with olive oil then add shrimp heads, until the light brown colour then add white wine reduce for 1 minute, then add canned tomato and red chili.
5. Add 2 litres water, fish bone and parsley stalks, simmer for 30 minutes then strain, then season with salt and pepper to be the base stock of tomato seafood pot.
6. Set up seafood platter, vegetable and tomato seafood pot, with pesto sauce on the side.

無骨牛小排附蘑菇白蘭地醬
Slices Boneless Beef Short Ribs with Mushroom Brandy Sauce

難易度 ★★　　　　　時間 90 分

材料 Ingredients	6 片	去骨牛小排	slices boneless beef short ribs
	30g	奶油	butter
	15g	麵粉	flour
	125cc	牛高湯	beef stock
	30cc	白蘭地	brandy
	150g	蘑菇	mushrooms
	45g	糖	sugar
	30cc	紅酒醋	red wine vinegar
	20g	蒜片	crispy garlic
洋蔥醃料 Onion Marinated Sauce	190cc	紅酒	red wine
	1/2 顆	洋蔥絲	onion, julienne
	30g	紅蔥頭碎	shallot, chopped
	1 片	月桂葉	bay leaf
	1 支	巴西利	parsley stalk
	適量 T.T.	鹽、黑胡椒碎	salt, chopped black pepper corn

作法 Method

1. 將洋蔥醃料拌勻，醃漬約 1 小時。
2. 將醋漬洋蔥煮滾，加入糖及紅酒醋，濃縮收汁為酒漬洋蔥。
3. 拌炒蘑菇至上色，用鹽、胡椒調味備用。
4. 用 15g 的奶油將麵粉炒香，加入牛高湯及醃汁攪拌至無顆粒，加入蘑菇片 濃縮至濃稠狀後再加入白蘭地及鹽、胡椒調味。
5. 將牛小排用鹽、胡椒調味，並用少許奶油至上色，至 5 分熟。
6. 上盤時附上酒漬洋蔥、蒜片及蘑菇白蘭地醬即可。

---

1. Mix well the onion marinade and pickle for about one hour.
2. Simmer the pickled onions and bring to boil. Add sugar and red wine vinegar, then reduce to be the red-wine macerated onions.
3. Sautéed mushrooms until the light brown colour and season with salt and pepper.
4. Make roux with butter and flour then add beef stock and stirring until thicken, then add mushroom slices and brandy and season it to be mushroom sauce.
5. Season beef short ribs with salt and pepper. Pan-fry with butter until it is medium.
6. Serve with red-wine marinated onions, crispy garlic and brandy mushroom sauce.

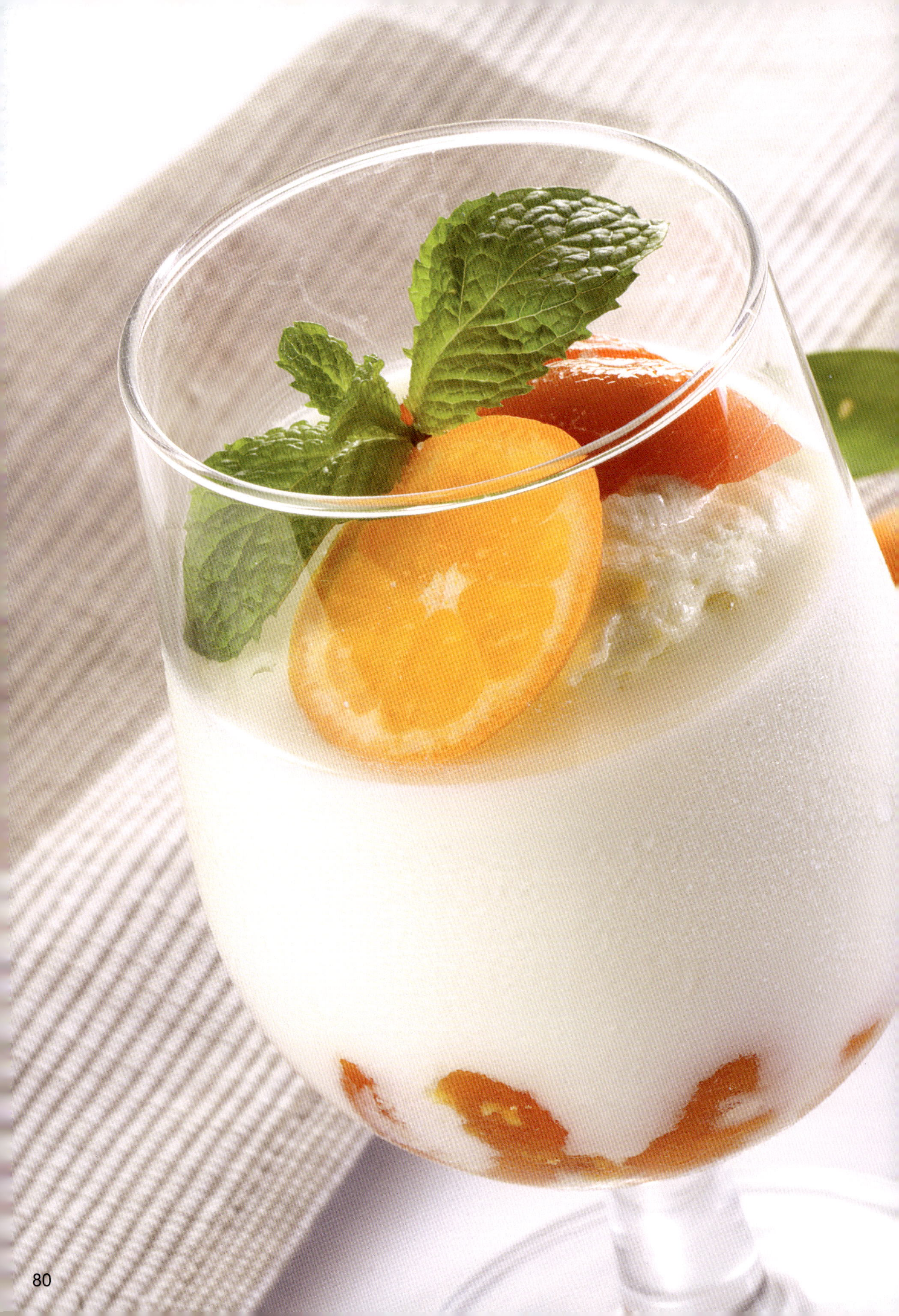

金桔奶酪
Kumquat Panna Cotta

難易度 ★　　　　時間 90 分

奶酪 材料
Ingredients of panna cotta

125cc	牛奶	milk
45g	糖	sugar
2 片	吉利丁片（泡冰水過濾備用）	gelatin (soaked iced water until soft)
30g	白巧克力，隔水加熱融化	white chocolate, melted.
250cc	原味優格	plain yogurt
125cc	無糖鮮奶油	fresh cream

奶酪 作法
Method of panna cotta

1. 用小火將糖溶入牛奶中，加入吉利丁片至完全溶化離火，再加入融化的白巧克力、原味優格及無糖鮮奶油攪拌均勻。
2. 將金桔餡放入香檳杯底，再加入奶酪。
3. 放入冰箱以 5℃ 冷藏約 1 小時至凝固，再用金桔及薄荷葉裝飾即可。

1. Melt sugar inside milk over the hot water then add gelatin until it is totally melted, then add white chocolate, yogurt and fresh cream then mix well.
2. Put kumquat stuffing on the bottom of the champagne glass then add panna cotta liquid.
3. Keep it in the fridge to be chilled at 5 ℃ for about 1 hour until done then garnish with kumquat and mint leaf.

金桔餡 材料
Ingredients of kumquat stuffing

6 顆	切片金桔	sliced kumquats
300cc	柳橙原汁	orange juice
125g	糖	sugar
3g	肉桂粉	cinnamon powder
5cc	新鮮檸檬汁	fresh lemon juice

金桔餡 作法
Method of kumquat stuffing

1. 將所有材料放入鍋中慢煮約 15 分鐘至金桔軟化，再濃縮至濃稠狀。

1. Simmer all ingredients until kumquats are softer, and then reduce it until it becomes thicken.

中國新年套餐二

Breakfast Set Menu 早餐 套餐

French Toast
法國吐司

Carrot Revitalizer and Hot Chocolat Coffee Latte
活力蘿蔔汁與熱巧克力咖啡拿鐵

Bircher Muesli
水果燕麥片

Ham and Cheese Omelet
火腿乳酪綜合蛋捲

Yogurt Potato Salad with Pan fried Sea bass
優格馬鈴薯沙拉鱸魚排

奧黛麗赫本（Audrey Hepburn），臉貼著第凡內（Tiffany）這個位於紐約第五街大道上，首屈一指的名貴珠寶店櫥窗，一邊吃著牛角麵包、喝著熱咖啡，一邊以艷羨的眼光望著第凡內中的一切，在第凡內早餐（Breakfast at Tiffany）的場景中，是那麼的雍容華貴。

但忙碌的現代人似乎都忽略了一天中最重要的第一餐，它是活力的泉源，每天只需花費少許的時間，即可做出豐盛營養又簡單的早餐。

Audrey Hepburn was staying close to the top valuable jewelry store which was on the Fifth Street in New York, she who was eating croissant, drinking hot coffee and was staring at everything inside the windows with envy sight. This scene is so elegant and poised in Breakfast at Tiffany. However, people seams to forget about the most important meal which is the most important meal of a day, it is the fountainhead of energy. You can spend just little time making a simple and nutrious breakfast.

法國吐司
French Toast

難易度 ★　　　　時間 20 分

材料 / Ingredients

1/2 條	法國麵包	baguett
100g	奶油	butter
30g	糖粉	icing sugar
A 500cc	牛奶	milk
2 顆	蛋	eggs
10g	糖	sugar
適量	肉桂粉、香草精、鹽	T.T. cinnamon powder, vanilla extract, salt

作法 / Method

1. 材料 A 攪拌均勻備用。
2. 將法國吐司切片沾上材料 A，以奶油煎成金黃色，撒上糖粉，再以草莓裝飾即可。食用時可附上楓糖醬。

1. Mix up equally the ingredients A.
2. Slice baguett and soak on ingredients A then pan-fry with butter to be golden brown colour, spread icing sugar garnish with strawberry.

早餐套餐

活力蘿蔔汁與熱巧克力咖啡拿鐵
Carrot Revitalizer and Hot Chocolat Coffee Latte

難易度 ★　　　　時間 15 分

蘿蔔汁 材料
Ingredients of carrot juice

2 條	紅蘿蔔	carrots
1 顆	蘋果	apple
4 顆	柳橙	oranges

巧克力拿鐵 材料
Ingredients of chocolate latte

1000cc	熱咖啡	hot coffee
60g	巧克力糖醬	chocolate sauce
少許	巧克力咖啡豆	few chocolate coffee beans
200cc	打發鮮奶油	whipped cream

蘿蔔汁 作法
Method of carrot juice

1. 將紅蘿蔔洗淨，去皮，切條；蘋果及柳橙去皮切塊。
2. 將紅蘿蔔、蘋果及柳橙用榨汁器壓成果汁，倒入杯中再用西芹裝飾即可。

1. Clean and peeled carrot, apples and oranges then cut to be big cubes.
2. Press carrot, apple and oranges with juicer to be juice garnish with celery.

巧克力咖啡拿鐵 作法
Method of chocolate latte

1. 將 10g 巧克力醬溶入 200cc 的熱咖啡中，攪拌均勻。
2. 擠上打發鮮奶油，再用巧克力糖醬及巧克力咖啡豆裝飾即可。

1. Melted 10g chocolate sauce into 200cc hot coffee then mix well.
2. Garnish with whipped cream, chocolate sauce and chocolate coffee beans.

水果燕麥片
Bircher Muesli

難易度　★　　　　時間 25 分

材料 Ingredients

6 顆	新鮮草莓	strawberries	
1 顆	奇異果	kiwifruit	
A 500cc	牛奶	milk	
200g	桂格大燕麥片	oatmeal	
30g	葡萄乾	raisin	
40g	杏仁片	almond slices	
40g	核桃	walnut	

B 125cc	無糖鮮奶油打發	whipped cream
20cc	蜂蜜	honey
1 顆	大紅蘋果	apple

作法 Method

1. 杏仁片、核桃先烤過，核桃切碎。材料A混合均勻，蓋上保鮮膜，放入冰箱冷藏至少12小時。
2. 蘋果去皮切絲；草莓及奇異果切丁備用。
3. 將浸泡的材料A拌入材料B拌勻，再用草莓及奇異果裝飾即可。

---

1. Toast almond slices and walnuts, and then chop walnuts. Mix up equally the ingredients A then cover with wrap and keep fridge for 12 hours.
2. Peeled of apple and cut it to be julienne; cut strawberry and kiwifruit to be dice.
3. Mix up the ingredients A with ingredients B then garnish with strawberry and kiwifruit.

早餐套餐

火腿乳酪綜合蛋捲
Ham and Cheese Omelet

難易度 ★★　　　　　　　時間 20 分

材料	12 顆	蛋	eggs
Ingredients	4 朵	新鮮蘑菇	mushrooms
	4 片	火腿片	sliced ham
	4 片	起司片	pc cheese
	120cc	沙拉油	oil
	20g	洋蔥碎	chopped onion
	適量	鹽、胡椒	T.T. salt, pepper
	8 顆	小蕃茄	cherry tomatoes

作法
Method

1. 將蛋打散後過濾；火腿切絲；蘑菇切片；蔥切碎備用。
2. 將所有材料除起司外炒香，加入蛋液拌炒至5分熟成半凝固蛋餅狀（3顆蛋可做1份），再加入起司絲，捲起成半月型煎至兩面金黃，待起司融化後加小蕃茄裝飾即可。

1. Beat eggs then strain it; cut ham to be julienne; cut mushrooms to be slices; chop spring onions.
2. Saut?all ingredients except cheese then add egg and until medium (3 eggs for one portion), then add cheese and roll it to be half moon shape and both sides are golden brown, then garnish with cherry tomatoes.

88

優格馬鈴薯沙拉鱸魚排
Yogurt Potato Salad with Pan fried Sea bass

難易度 ★★　　　　時間 45 分

材料 Ingredients

250g	無糖優格	sugar-free yogurt
400g	水煮馬鈴薯	boiled potatoes
50g	紅蔥頭(炒香)	red onion (sautéed)
10cc	檸檬汁	lemon juice
10g	巴西利(切碎)	parsley (chopped)
10g	青蔥(蔥花)	shallots (scallions)
150g	菠菜葉(燙煮)	spinach leaves (scalded)
1顆	蒜頭	garlic
50cc	橄欖油	olive oil
4片	鱸魚(片/280g)	seabass (fillet / 280g)
10g	麵粉	flour
適量	鹽、胡椒	T.T. salt and pepper

配菜 Garnish

烤甜椒
150g	紅甜椒	red bell pepper
150g	黃甜椒	yellow bell peppe

乾烤後切絲用鹽、胡椒調味　After roast, season with salt and pepper.

作法 Method

1. 將熱馬鈴薯壓成泥，依序拌入無糖優格、炒香紅蔥頭炒香、檸檬汁、巴西利及蔥花拌均勻，並用鹽、胡椒調味。
2. 再拌入 150g 菠菜絲。
3. 將另一半菠菜與蒜頭加入 30cc 橄欖油打成醬汁並用鹽、胡椒調味。
4. 將鱸魚片對切共 8 片，並用鹽、胡椒調味，拍上麵粉煎成兩面呈金黃色至熟，
5. 裝飾：醬汁(作法 3)與優格馬鈴薯沙拉做基底，上方放鱸魚排，烤甜椒絲即可。

---

1. Press the hot potato to be mashed. Add sugar-free yogurt, sautéed red onion, lemon juice, parsley, and scallions sequentially. Stir well. And season with salt and pepper to taste.
2. Mix in 150 grams of spinach wire.
3. Add 30cc of olive oil into the other half of the spinach and garlic. Stir well until smooth. Season with salt, pepper.
4. Cut sea bass fillets in half into 8 slices, and season with salt and pepper, then dust with flour and pan fry until light golden brown on both sides.
5. Garnish: Use sauces (method 3) and yogurt potato salad as the base, and top with sea bass fillets and roasted sweet pepper, and serve.

Brunch Set Menu 早午餐 套餐

Pancake with Maple Syrup
美式鬆餅附楓糖醬

Sweet Corn Chowder in French White Bread
麵包甜玉米巧達湯

Baked Egg with Creamy Spinach
焗奶油菠菜烘蛋

Fried Breakfast Sausage, Bacon and Roll Ham
香煎早餐腸、培根及火腿

Maple Leaf Frappe
楓香奶昔

一般早餐分為歐陸式早餐「Continental Breakfast」，只包括各類的麵包及飲料，較為簡單；另一為美式早餐「America Breakfast」，其內容豐盛，增加了各類的熱菜、麥片、蛋類等，多樣的選擇是目前一般大眾較喜歡的類型。假日總是偷閒多睡點，演變成早餐及午餐結合的「Brunch」早午餐，更成為現代人放鬆自己的另一種享受。

Audrey Hepburn was staying close to the top valuable jewelry store which was on the Fifth Street in New York, she who was eating croissant, drinking hot coffee and was staring at everything inside the windows with envy sight. This scene is so elegant and poised in Breakfast at Tiffany. However, people seams to forget about the most important meal which is the most important meal of a day, it is the fountainhead of energy. You can spend just little time making a simple and nutrious breakfast.

美式鬆餅附楓糖醬
Pancake with Maple Syrup

早午餐套餐

難易度 ★　　　　　時間 15 分

材料
Ingredients

A	適量	奶油	T.T. butter	B 250cc	牛奶	milk
	200g	高筋麵粉	bread flour	30cc	沙拉油	oil
	100g	糖	sugar	2 顆	蛋黃	Egg yolk
	10g	泡打粉	baking powder	C 2 顆	蛋白,打發	Egg white, whipping
	2g	鹽	salt			

作法
Method

1. 將材料 A 拌勻，加入材料 B 用打蛋器攪拌至無顆粒狀再拌入材料 C 打發蛋白，做成鬆餅麵糊。
2. 取一平底鍋轉中小火，塗上一層薄奶油後，加入 15 g 麵鬆餅糊煎至兩面呈金黃色即為鬆餅。
3. 食用時可附上楓糖醬及新鮮水果裝飾。

1. Mix up equally the ingredients A then add ingredients B and whip until paste come to smooth and add ingredients C.
2. Use moderate heat, then add 15g paste with butter and fry until the both sides are golden brown colour.
3. Serve with maple syrup and fruit.

材料 Ingredients

1 支	白玉米	white corn	
2 支	黃玉米	sweet corns	
60g	奶油	butter	
120g	洋蔥碎	chopped onion	
2 支	西芹碎	chopped celeries	
500g	洋芋丁	diced potato	
500cc	雞高湯	chicken stock	
250cc	水	water	
5g	百里香	thyme	
250cc	無糖鮮奶油	cream	
適量	鹽、胡椒	T.T. salt, pepper	
10g	蝦夷蔥或青蔥	chives or spring onion	
4 顆	中型法國圓麵包（於 1/4 頂部平切為湯蓋，麵包挖空為湯碗）		

middle size French round breads
(cut 1/4 from the top to be the soup cover, then excavate the bread to be the soup bowl)

麵包甜玉米巧達湯
Sweet Corn Chowder in French White Bread

難易度 ★　　　　　時間 40 分

作法
Method

1. 取白、黃玉米粒備用,預留 1/2 杯裝飾用。
2. 用奶油將西芹、洋蔥及玉米粒炒香,再加入洋芋、雞高湯、玉米梗及水,煮至沸騰,加入百里香,以小火燜煮 20 分鐘後,將玉米梗取出。
3. 將步驟 2 放入果汁機打散至無顆粒,加入無糖鮮奶油及裝飾玉米粒,再煮至沸騰,放入少許鹽、胡椒調味。
4. 將麵包放入 180℃ 烤箱烤約 1 分鐘。
5. 將湯裝入麵包碗中,用蝦夷蔥裝飾,蓋上麵包蓋即可。

1. Take off white and yellow corn, and keep 1/2 cup for garnish.
2. Sauté celery, onion and corn with butter then add potato, chicken stock, corn stalks, thyme and water and bring to simmer for 20 minutes, then take out the corn stalks.
3. Pureé the soup then strain it, then add cream, corns and bring to boil, season to taste
4. Bake the bread in oven at 180℃ for about 1 minute.
5. Pour soup into the bread bowl and garnish with chives.

焗奶油菠菜烘蛋
Baked Egg with Creamy Spinach

難易度 ★★　　　　時間 40 分

材料 Ingredients

800g	菠菜	spinach
4 顆	蛋	eggs
80g	起司絲	grated cheese
A 500cc	牛奶	milk
45g	中筋麵粉	flour
45g	奶油	butter
100g	奶油乳酪	cream cheese
適量	鹽、胡椒	T.T. salt, pepper

作法 Method

1. 烤箱預熱 180℃；將菠菜洗淨切段，燙煮後沖冷水備用。
2. 將材料 A 的麵粉及奶油炒成麵糊後，加入 500cc 牛奶攪拌均勻至無顆粒，煮滾後再拌入奶油起司至融化，加入鹽、胡椒調味後即為奶油醬。
3. 將菠菜拌入奶油醬中，倒入焗盤內，加入蛋及起司絲，放入烤箱烤 15 分鐘至呈金黃色即可。

---

1. Preheat oven to 180℃; clean and cut spinach to be strips, blanch.
2. Make roux from butter and flour then add milk and stirring until smooth and add cream cheese mix well until cheese melted and season to be cream sauce.
3. Mix spinach into cream sauce, then place in the baking plate, add egg and grated cheese and then bake for 15 minutes until done.

香煎早餐腸、培根及火腿
Fried Breakfast Sausage , Bacon and Roll Ham

難易度 ★　　　　　　時間 15 分

材料 Ingredients

8 片	培根	pc bacon
8 條	早餐香腸	breakfast sausages
8 片	火腿片	slice ham
30g	奶油	butter

作法 Method

1. 將培根及火腿片煎至兩面呈金黃色即可。
2. 將早餐香腸放入熱水燙煮 1 分鐘，再以奶油煎上色即可。

1. Pan-fry bacon and ham until the both sides are golden brown.
2. Blanch breakfast sausages for 1 minute then pan-fry with butter until done.

楓香奶昔
Maple Leaf Frappe

難易度 ★　　　　時間 15 分

材料　　　　　　　　　　　　　　　
Ingredients

120cc	楓糖醬	maple syrup
16 顆	冰塊	ice cubes
4 顆	香草冰淇淋（約 120g）	scoops of vanilla ice cream (about 120g)
500cc	牛奶	milk
4 顆	紅櫻桃	cherries

作法
Method

1. 將楓糖醬、冰塊及香草冰淇淋打成泥狀。
2. 加入牛奶，攪拌均勻。
3. 淋上少許楓糖醬及紅櫻桃裝飾即可。

---

1. Blend maple syrup, ice cubes and vanilla ice cream with blending until smooth.
2. Add milk and mix well.
3. Pour into glass garnish with maple syrup and cherry.

Afternoon Tea Menu 午茶休閒餐

Tuna and Avocado Salad
鮪魚酪梨沙拉

Chicken Pot Pie
乳酪雞肉派

Scottish Scones
英式比士吉

Ham and Leek Quiche
火腿蒜苗塔

Apple Muffin
蘋果馬芬蛋糕

在英國人的生活當中，下午茶不但具有社交意義，也為生活增添情趣。此菜單有英式口味的 Quiche 火腿蒜苗塔及 Scones 比士吉，還有美式口味的 ChickenPot Pie 乳酪雞肉派、鮪魚酪梨沙拉與 Muffin 馬芬蛋糕，是您在午茶休閒 Tea Break 的最佳餐點！

In British life, afternoon tea is not only a social activity but also a spicy to life. Ham and leek quiche and scones are British style and American chicken pot pie, Tuna and Avocado Salad and muffin are the best snack for your afternoon tea time.

鮪魚酪梨沙拉
Tuna and Avocado Salad

難易度 ★★　　　　　　時間 45 分

午茶休閒餐

材料 Ingredients				
A	100g	洋菇 (切片)	mushrooms (sliced)
	30cc	檸檬汁	lemon juice
	1g	鹽	salt
	1g	黑胡椒	black pepper
	200g	酪梨 (切 1/4 後切片)	avocado (cut 1/4 then sliced)
B	15cc	橄欖油	olive oil
	15cc	紅酒醋	red wine vinegar
	5g	鯷魚	anchovies
	3g	蒜頭 (切碎)	garlic (chopped)
C	300g	新鮮鮪魚 (切長條狀)	fresh tuna (cut into strips)
	3g	鹽	salt
	6g	黑胡椒碎	crushed black pepper
	2g	花椒 (切碎)	pepper (chopped)
	1g	白芝麻 (烤過)	white sesame seeds (toasted)
D	200g	紅蕃茄 (切角 1/8)	red tomatoes (cut into 1/8 on the diagonal)
	120g	紅洋蔥 (切片)	red onion (sliced)
	100g	黃甜椒 (切片)	yellow bell pepper (sliced)
	1/2 顆	檸檬 (去皮切薄片)	lemon (peeled and sliced)
	4 顆	黑橄欖	black olives
E	3g	巴西利 (切碎)	parsley (chopped)
	1 顆	水煮蛋 (切角 1/6)	boiled eggs (cut into 1/6 on the diagonal)
	30g	綠捲鬚生菜	frisée

作法 Method

1. 將洋菇拌入檸檬汁並用鹽及黑胡椒調味，充分拌均勻後加入酪梨片後拌勻醃製 5 分鐘備用；將鯷魚搗碎拌入蒜頭碎、橄欖油及紅酒醋並用打蛋拌攪均勻為醬汁備用。
2. 鹽、黑胡椒碎、花椒碎及白芝麻拌勻，並將鮪魚沾上調味料，乾煎至 4 個面上色，切 0.5 公分厚片。
3. 最後將材料 A 至 D 拌均勻上盤後再用水煮蛋裝飾並撒上巴西利碎及綠捲鬚生菜。

1. Stir together mushrooms and lemon juice, season with salt and black pepper, add avocado slices, mix well then marinate for 5 minutes and set aside; Pound the anchovies, add chopped garlic, olive oil, and red wine vinegar, stir well with a whisk to be the sauce. Set aside.
2. Mix well salt, crushed black pepper, chopped pepper, and white sesame seeds. Coat the tuna with seasonings, saute until brown on four sides, slice into 0.5 cm thick slices.
3. Mix evenly the ingredients A~D and place them on the plate, then garnish with boiled egg and sprinkle with chopped parsley and frisée.

乳酪雞肉派
Chicken Pot Pie

難易度 ★★　　時間 50 分

材料 Ingredients

	150g	雞胸（切 1.5 公分小丁）	chicken breast, dice
	30g	洋蔥末	chopped onion
	60g	冷凍三色蔬菜丁	frozen dice vegetable
	200cc	雞高湯	chicken stock
	100cc	牛奶	milk
	300g	鹹派皮（詳 P.103）	salty pie dough
	4 個、3 吋	塔模	models with 3 inches size
	適量	蛋液（蛋黃 1 顆加 15cc 水打散備用） T.T. egg wash(1 egg yolk and 15cc water)	
A	30g	麵粉	flour
	30g	奶油	butter
B	20g	起司絲	grated cheese
	5g	帕瑪起司粉	Parmesan cheese powder

作法 Method

1. 烤箱預熱至 190℃，雞胸肉燙熟，洋蔥、三色蔬菜丁炒香備用。
2. 將材料 A 用中小火炒成麵糊 ROUX，再加入高湯與牛奶攪拌至無顆粒奶油狀，續加入雞肉丁及蔬菜丁，待涼後再拌入材料 B 備用。
3. 將鹹派皮 成 8 張 5 吋 0.3 公分的圓形派皮。
4. 將派皮放入塔模中，填入餡料 8 分滿，在派皮四周擦上蛋液再蓋上另一張派皮，表面再擦上蛋液，用叉子在表面刺洞，送入烤箱烤 25 分鐘至呈金黃色即可。

備註：派皮刺洞可使餡料在烘烤時空氣可以排出，亦可將雞肉更換成綜合海鮮，做成海鮮派。

1. Preheat oven to 190℃, Blanch chicken breast and saut?onion and vegetable with butter.
2. Make roux from butter and flour, then add chicken stock and milk and stirring until smooth, then add chicken breast, vegetable and cheese from ingredient B.
3. Rolling dough to be 8 pieces round sheet, 5 inches each.
4. Put dough into model and put stuffing into it, egg wash on the edge of the dough then cover with another dough, egg wash on the surface again, then stab on the surface with a folk, Bake for 25 minutes until done.

英式比士吉
Scottish Scones

難易度 ★　　　　　　　　時間 40 分

材料 Ingredients

	30g	草莓果醬	strawberry jam
A	300g	低筋麵粉	cake flour
	60g	奶油	butter
	10g	泡打粉	baking powder
B	50cc	蘭姆酒	rum
	40g	葡萄乾	raisins
C	100cc	牛奶	milk
	50g	糖	sugar
	1 個	蛋	egg
	1g	鹽	salt

作法 Method

1. 烤箱預熱至 225℃，將材料 B 浸泡 30 分鐘以上，瀝乾備用。
2. 將材料 A 用手攪至奶油溶入麵粉成顆粒粉狀。
3. 再將材料 C 打散，拌入步驟 3 中揉 5 分鐘至表面光滑不黏手。
4. 將麵糰 成 2 公分厚，加入醃漬葡萄乾，再將麵糰對折約 4 公分厚度。
5. 用 2 吋圓型模型將麵糰蓋成數個圓形，蓋濕布醒約 15 分鐘，擦上蛋液，送入烤箱烤約 15 分鐘至呈金黃色即可，食用時可附上草莓果醬。

1. Preheat oven to 225℃, Soak rum and raisins for more than 30 minutes then strain it.
2. Mix up the ingredients A by hands until butter is meled into flour.
3. Add the ingredients C then toss and rub for 5 minutes until the surface is smooth.
4. Rolling the dough to be 2cm thick then add marinated raisin and then fold it to be 4cm thick.
5. Make round dough by 2 inches round model, cover rest with a wet cloth for about 15 minutes, and egg wash bake for 15 minutes until golden brown, Serve with strawberry jam or whipping cream.

午茶休閒餐

火腿蒜苗塔
Ham and Leek Quiche

難易度 ★★　　　　　時間 45 分

材料 Ingredients

	150g	鹹派皮	salty pie dough
	50g	起司絲	grated cheese
	適量	紅豆	T.T. red beans
	20g	奶油	butter
	4 個	派盤 3 吋	tart mold
A	50g	火腿	ham, dice
	40g	培根	bacon, dice
	1 支	青蒜	leek, dice
	1/4 顆	洋蔥碎	chopped onion
	2~3 朵	洋菇片	mishrooms, sliced
B	1 顆	蛋	egg
	65cc	牛奶	milk
	30cc	無糖鮮奶油	fresh cream
	適量	鹽、胡椒、豆蔻粉	T.T. salt, pepper, nutmeg powder

鹹派皮 材料 Ingredients of salty pie dough
(約 450g) (about 450g):

	80cc	冰水	iced water
A	175g	高筋麵粉	bread flour
	75g	低筋麵粉	cake flour
	125g	奶油	butter
	5g	鹽	salt

鹹派皮 做法 Method of salty pie dough

將材料 A 混合均勻，用手抓奶油與麵粉，使麵粉成米粒狀，再加入冰水拌勻成麵糰，放入冰箱冷藏備用。

Grab butter, salt and flour with hands and make it to be rice shape, then add iced water and mix well then keep in fridge.

作法 Method

1. 烤箱預溫 165℃，派盤塗上薄薄一層奶油。
2. 將塔皮桿成 0.4 公分厚度，放入派盤中，鋪上紅豆，放入烤箱以 165℃ 烤約 12 分鐘至熟，取出後去除紅豆，即成派皮。
3. 將材料 B 拌均勻、過濾備用。
4. 鍋熱加入奶油，用中小火將步驟 3 的培根炒香，再加入其他材料拌炒約 2 分鐘至上色，以鹽、胡椒調味，即成內餡。
5. 將內餡放入塔皮中加入起司絲，倒入蛋液約 8 分滿，入烤箱以 175℃ 烤約 20 分鐘至金黃色即可。

※ 加紅豆是為了防止在烘烤派皮時派皮的膨脹。

1. Preheat oven to 165℃, spread butter on mold.
2. Rolling the dough to be 0.4 thick and put in a mold, then spread red beans and bake 165℃ for 12 minutes until it is cooked, take it out and take off red beans to be tart crust.
3. Whip and mix well from ingredients B and strain.
4. Sautéd the bacons with butter then add other ingredients A for 2 minutes until brown colour, and season.
5. Put the stuffing into the tart crust then add cheese and then pour egg liquid for about 8/10 of it, bake 175℃ for about 20 minutes until done.

※ Add red bean inside the tart are for avoiding the tart crust inflates.

蘋果馬芬蛋糕
Apple Muffin

難易度 ★　　　時間 40 分

材料 Ingredients

	數量	中文	English
	1 顆	蘋果	apple
	70g	核桃	walnut
A	100g	糖	sugar
	90cc	蔬菜油	vegetable oil
	65cc	鮮奶	milk
	1 顆	蛋	egg
	5g	香草精	vanilla extract
B	185g	中筋麵粉	half white flour
	5g	泡打粉	baking powder
	1g	鹽	salt

作法 Method

1. 烤箱預熱 180℃。
2. 蘋果去皮切小丁；核桃烤上色亦切丁備用。
3. 將材料 B 混合均勻，材料 A 打散，拌入材料 B 攪拌均勻後再加入蘋果及核桃。
4. 將馬芬麵糊裝入馬芬紙杯中，烤約 25 分鐘呈金黃色即可。

---

1. Preheat oven to 180℃.
2. Peeled and dice the apple, toast walnuts to be light brown colour and chop.
3. Mix well the ingredients A and B, then fold the apple and walnut.
4. Place muffin paste into muffin cup and bake for about 25 minutes until done.

Great DVD Set Menu
DVD 風味 休閒餐

Clam Chowder
蛤蜊巧達湯

Cheesy Grilled Mussels
乳酪烤淡菜

Seafood Pizza
海鮮披薩

Grass Shrimp and Avocado Sandwich
鮮蝦酪梨三明治

Chicken Fingers and Cheese Sticks with Tartar Sauce
香酥雞肉條與炸起司條附塔塔醬

Banana Ice Cream Shake
香蕉奶昔

你是否曾邀約三五好友到家中，挑支優質的 DVD 影片一起觀賞？不用再為了要吃些什麼而煩心了，告別泡麵、爆米花、玉米片等零食的日子吧！
這是一套簡易且適合看片子的菜單，一起與好友分享這套有料的 DVD 大餐吧！

Have you ever picked up a good DVD and invited some good friends to your house then enjoyed together?
Stop worrying about what to eat, just say goodbye to instant noodles, popcorn and chips and so on!
This is a simple menu and suitable for watching film, which enables you to enjoy with good friends!

蛤蜊巧達湯
Clam Chowder

難易度 ★★　　　　　時間 80 分

材料 Ingredients

	350g	蛤蠣	clams
	1 支	西芹	celery
	1/2 顆	洋蔥	onion
	1 顆	洋芋	potato
	5g	巴西利切碎	chopped parsley
	60g	奶油	butter
	180cc	無糖鮮奶油	fresh cream
	500cc	牛奶	milk
	30cc	茴香酒或白酒	pernod or white wine
A	50g	奶油	butter
	50g	麵粉	flour
B	2.5g	百里香	thyme
	1 片	月桂葉	bay leaf
	少許	鹽、胡椒	T.T. salt, pepper
C	60g	蝦仁	shrimps
	60g	花枝	squid

作法 Method

1. 蛤蜊用鹽水吐沙約 1 小時後洗淨；洋蔥、西芹切小丁；洋芋去皮切丁泡水備用。
2. 煮 1500cc 水將蛤蜊燙至開口，取肉，湯汁濃縮成 1000cc，過濾備用。
3. 以奶油炒洋蔥、西芹至軟化；洋芋用水煮 8 分鐘至熟備用；將材料 C 切丁燙煮備用。
4. 將材料 A 炒成淡黃色麵糊，慢慢加入蛤蜊湯及牛奶，攪拌均勻至無顆粒狀，再依續加入洋蔥、西芹、洋芋、材料 B 及材料 C 煮 15 分鐘。
5. 加入鮮奶油調整濃稠度，再加入茴香酒（或白酒）、鹽、胡椒調味即可。
6. 最後以香菜葉裝飾。

---

1. Make clams through out sends by salt water for about 1 hour then clean; cut onion and celeries to be dice; peeled potatoes and dice.
2. Boil 1500cc water and blanch clams until the shells are opened, take the meat out and reduce the stock to be 1000cc then strain it.
3. Sauté onion and celery with butter until soft; boil potatoes until soft; dice the seafood and blanch.
4. Make roux form butter and flour, then add clam stock and milk, stirring until smooth, and add onion, celery, herb and seafood and simmer for 15 minutes.
5. Add fresh cream to make it to be thicken, then add pernod (or white wine), salt and pepper to taste.
6. Garnish with coriander.

乳酪烤淡菜
Cheesy Grilled Mussels

難易度 ★　　　　時間 25 分

材料 Ingredients

份量	中文	English
500cc	水	water
12 個	淡菜	mussels
2 顆	蒜頭碎	chopped garlic
5g	香菜碎	chopped coriander
1 支	辣椒，去籽切碎	chili seeded and chop
1 顆	檸檬，切角	lemon wages
5g	檸檬皮碎	lemon zest
30g	帕瑪森起司粉	Parmesan Cheese Powder
250g	麵包粉	bread crumbs
60g	融化奶油	melted butter
適量	鹽、胡椒	T.T. salt, pepper

作法 Method

1. 烤箱預溫 180℃。
2. 將淡菜用滾水燙煮約 3 分鐘後，過濾，取出淡菜肉，清洗淡菜殼，再將其淡菜肉放入殼中。
3. 香料麵包粉：將蒜頭、香菜、辣椒、檸檬皮、起司粉及麵包粉拌均勻，再加入融化奶油、適量的鹽、胡椒調味。
4. 將香料麵包粉蓋在淡菜上，送入烤箱烤 5 分鐘至香料麵包粉呈金黃色即可。
5. 附上檸檬角裝飾。

---

1. Preheat oven to 180℃.
2. Blanch mussels then take out the mussel meat, clean the shells and then put the mussel meat back into shells.
3. Herbal bread crumbs: mix garlic, coriander, chili, lemon zest, parmesan cheese and bread crumbs then add melted butter, salt and pepper.
4. Cover the herbal bread on the mussels, bake for 5 minutes until golden brown.
5. Garnish with lemon wages.

海鮮披薩
Seafood Pizza

難易度 ★★　　　　　　　　時間 60 分

材料 Ingredients

皮	250g	高筋麵粉	bread flour
	150cc	溫水	warm water
	15g / 5g	新鮮酵母 / 乾酵母	fresh yeast/dry yeast
	15g	橄欖油	olive oil
	8g	糖	sugar
餡	100g	披薩起司絲	grate cheese
	100g	蛤蜊（湯煮取肉）	clam meat
	150g	蝦仁	shrimps
	120g	花枝（切圈，湯煮備用）	squid (cut to be ring, blanch)
	3 顆	洋菇	mushroom, sliced
	1/4 顆	洋蔥	onion julienne
	5 片	九層塔	pieces of basil
醬汁	125g	去皮蕃茄碎	peeled tomato and chop
	30g	洋蔥碎	chopped onion
	2.5g	奧勒崗	oregano
	適量	鹽、胡椒	T.T. salt, pepper

作法 Method

1. 披薩醬汁：將洋蔥炒香，加入去皮蕃茄、奧勒崗及適量鹽、胡椒調味；烤箱預熱 200℃。
2. 將酵母、鹽、水混勻後再加麵粉及奶油揉至不沾手，分成 2 個，蓋上濕布，醒麵 30~40 分鐘。
3. 將麵糰揉成圓形，用叉子刺洞，放入烤箱以 210℃ 烤 8 分鐘至呈淡黃色，取出備用。
4. 塗上披薩醬汁，加上綜合海鮮、洋菇、洋蔥、九層塔及起司絲，入烤箱以 200℃ 烤約 12 分鐘至金黃色即可。

---

1. Pizza sauce : sauté onion then add peeled tomato, oregano, salt and pepper to taste. Preheat oven to 200℃.
2. Mix up equally yeast, sugar and water then add flour and olive oil and rub it until smooth and elastic, then divide it to be two each, cover with wet cloth to ferment for 30~40 minutes.
3. Rolling dough to be round and stab with folk, then bake it in oven at 210℃ for 8 minutes until the light brown colour.
4. Spread pizza sauce, add assorted seafood, mushrooms, onion, basil and cheese, bake for about 12 minutes until done.

DVD 風味休閒餐

鮮蝦酪梨三明治
Grass Shrimp and Avocado Sandwich

難易度 ★★　　　　　　時間 20 分

材料 Ingredients

6 片 slices	全麥吐司	whole-wheat toast
16 隻	草蝦 (320g)	pc grass shrimp (320g)
1 顆	紅蕃茄 (切片)	red tomato (sliced)
4 片	乳酪片	cheese slices
100g (1/2 顆)	酪梨 (去籽切片)	(1/2) avocado (seed removed, sliced)
60g (8 片)	生菜葉	(8) lettuce
40g	酸黃瓜 (切片)	pickled cucumber (sliced)

醬料 Sauces

10 g	美乃滋	mayonnaise
30g	蕃茄醬	ketchup
5g	黃芥末醬	yellow mustard
5cc	檸檬汁	lemon juice
5g	蒜頭碎	crushed garlic

燙蝦高湯 Broth for boiling shrimp

(1/2 洋蔥塊、1/2 紅蘿蔔塊、1 支西芹塊、1 片月桂葉、5cc 白醋及 1 公升水)
(1/2 onion pieces, 1/2 carrot pieces, 1 piece celery, 1 bay leaf, 5cc white vinegar, and 1 liter of water)

作法 Method

1. 草蝦洗淨，用牙籤從蝦 2 至 3 節中插入，去除腸泥；準備燙蝦高湯材料煮滾後轉小火煮 25 分鐘過濾備用。
2. 將草蝦放入蔬菜高湯中燙約 90 秒，泡入冰水中冰鎮，去蝦殼備用。
3. 將醬料攪拌均勻。
4. 吐司烤上色 (3 片為兩份) 塗上醬料；
 分別夾上生菜、乳酪片、酪梨、酸黃瓜、草蝦為一層；
 生菜、乳酪片、紅蕃茄、酪梨、草蝦為另外一層，對切成 4 個三角形即可。

---

1. Wash the shrimp. Pull the vein out by inserting a toothpick between section 2 and 3 of the shrimp. Prepare the ingredients for boiling shrimp, bring to boil, reduce heat and simmer for 25 minutes, then filter.
2. Put the shrimp into the vegetable broth for about 90 seconds, soaked in ice water, remove the shell and set aside.
3. Stir the sauces evenly.
4. Brown the toast (3 slices yield 2 servings), and spread the sauce.
 Put lettuce, cheese slices, avocado, pickled cucumber, and grass shrimp to be the first layer.
 Put lettuce, cheese slices, red tomatoes, avocado, and grass shrimp to be another layer. Then cut each sandwich into quarters on the diagonal, and serve.

香酥雞肉條與炸起司條 附塔塔醬
Chicken Fingers and Cheese Sticks with Tartar Sauce

難易度 ★　　　　　　時間 45 分

塔塔醬 材料　Ingredients of Tartar Sauce

250g	美乃滋	mayonnaise
10g	法式芥末	Dijon mustard
45g	碎酸黃瓜	chopped pickled cucumber
1 顆	水煮蛋碎	chopped boiled egg
30g	洋蔥碎	chopped onion
5g	巴西利碎	chopped parsley
15cc	檸檬汁	lemon juice
少許	鹽、胡椒	T.T. salt, pepper
10 顆	酸豆碎	chopped capers

塔塔醬 作法　Method of Tartar Sauce

1. 將美乃滋及所有的材料混合均勻即可。

1. Mix up equally mayonnaise and all ingredients, season salt and pepper to taste.

材料 Ingredients

	1/2 顆	檸檬角	lemon wages
A	2 片	雞胸肉切條 1.5×1.5×8cm	pieces of chicken breast, cut strips
	10g	匈牙利紅椒粉	paprika
	適量	鹽、胡椒	T.T. salt, pepper
	1/2 顆	檸檬汁	lemon juice
B	120g	巧達起司，切條 0.8×0.8×8cm	chowder cheese, cut strips
C	沾裹 Breading：200g	麵粉	flour
	2 顆	蛋，打散	eggs, beated
	200g	麵包粉	bread crumbs

作法 Method

1. 將材料 A 醃製 15 分鐘備用。
2. 將雞肉條及巧達起司依麵粉、蛋液、麵包粉的順序沾裹。
3. 備一油鍋，待鍋中油溫至 180℃，分別放入炸雞肉條及巧達起司，炸至呈金黃色即可。
4. 附上檸檬角及塔塔醬。

---

1. Marinate chicken from ingredients A for 15 minutes.
2. Breading chicken strips and chowder cheese on flour, egg liquid and bread crumbs.
3. Preheat deep fry oil to 180℃, put chicken and cheese individual and deep-fry them until done.
4. Serve with lemon wages and tartar sauce.

香蕉奶昔
Banana Ice Cream Shake

難易度 ★   時間 15 分

材料 / Ingredients

A	250cc	鳳梨汁	pineapple juice
	250cc	椰奶	coconut milk
	2 條	香蕉	bananas
	4 球	香草冰淇淋（1 球 約 30g）	scoops of vanilla ice cream (1 scoop is about 30g)
	250cc	牛奶	milk
	適量	冰塊	ices
B	1 顆	奇異果	kiwifruit
	1 條	香蕉	banana

作法 / Method

1. 將材料 A 用果汁機打均勻。
2. 用奇異果及香蕉裝飾即可。

1. Blend ingredients A by blender.
2. Garnish with kiwifruit and banana.

Sandwich Set Menu 三明治套餐

Cream of Cauliflower Soup
奶油白花菜湯

Tuna Submarine
鮪魚潛水艇三明治

Club Sandwich
總匯三明治

Egg Salad Sandwich
蛋沙拉三明治

Summer Fruit Salad Sandwich
夏日水果沙拉三明治

　　大多數的人都認為三明治只是一種簡單、可果腹的點心，其實，只要運用巧思及各類食材，它也可以呈現出多種不同的風貌。三明治（Sandwich）始於英國，因貴族們熱衷於玩牌，又怕耽誤正餐，所以將喜歡的烤肉夾在麵包中，邊吃邊玩。爾後傳至法國，盛行於美國。因材料豐富、作法簡單、變化性大，且又方便攜帶，給忙碌的現代人莫大的便利，所以一直流傳至今，廣受大眾的歡迎。

　　這套套餐中包含了熱的三明治──總匯三明治，冷夾心三明治──蛋沙拉三明治，以法國麵包做成的鮪魚潛水艇三明治及甜點──夏日水果沙拉三明治，再搭配奶油花菜濃湯。現在，就一起來享受三明治套餐吧！

Most people consider that sandwiches are a kind of simple snack; actually, it could be different kinds of style if you try to create and use other food. Sandwiches originated from British, for nobles liked to play pokers but were afraid of meals delayed so that they put meat inside breads, eating and playing simultaneously. Then the food went to France and was in vogue in United States. Sandwiches contain abundant ingredients. There are simple, changeable and easy to carry with, therefore, they are always popular.

This set menu contains hot sandwich-club sandwich, cold with filling sandwich-salad sandwich with egg,Tuna submarine sandwich is made of French bread and dessert-summer fruit salad sandwich with creamy cauliflower soup. Now let's enjoy sandwich set!

奶油白花菜湯
Cream of Cauliflower Soup

難易度 ★★　　　時間 60 分

材料 Ingredients

150g	奶油	butter
120g	中筋麵粉	A.P. flour
120g	洋蔥	onion
60g	青蒜	leek
60g	西芹	celery
1 1/2 公升	雞高湯	litre chicken stock
1 顆	白花菜	cauliflower（約 600g）
適量	鹽、胡椒	T.T. salt, pepper
250cc	無糖鮮奶油	fresh cream

香料包 Spices bag

1 片	月桂葉	bay leaf
5 顆	白胡椒粒	white pepper corns
1 支	百里香葉	thyme 做成香料包

作法 Method

1. 花菜預留 1 杯的量（花的部分）燙煮，沖冷水備用，將其餘的花菜切碎。
2. 用中小火以奶油炒香洋蔥、青蒜、西芹及白花菜碎至軟，再加入中筋麵粉拌炒約 3 分鐘。
3. 加雞高湯攪拌至無麵粉顆粒，再加入香料包以小火煮 45 分鐘。
4. 過濾湯渣，再加入鮮奶油及裝飾的白花菜煮滾，以鹽、胡椒調味即可。

1. Keep 1 cup of cauliflower and blanch, and then chop the rest of cauliflower.
2. Sauté onion, leek, celery and chopped cauliflower with butter until soft then add flour and keep frying for about 3 minutes.
3. Add chicken stock and stirring until no granule and smooth then add spicy bag, simmer for 45 minutes.
4. Strain then and add fresh cream and cauliflower garnish bring to boil, and season with salt and pepper to taste.

鮪魚潛水艇三明治
Tuna Submarine

難易度 ★　　　　時間 20 分

材料 Ingredients

	1 條	法國麵包	loaf baguett
	30g	美乃滋	mayonnaise
A	1 罐	鮪魚罐	tuna can
	1/2 顆	洋蔥	onion
	6 顆	黑橄欖	black olives
	1 條	小黃瓜	cucumber
	1/2 個	紅椒	red bell pepper
	1/2 個	青椒	green pepper
	30g	美乃滋	mayonnaise
B	1/4 個	美生菜	lettuce
	2 個	蕃茄	tomato
	15g	法式芥末醬	Dijon mustard
	4 片	起司片	cheese slices

作法 Method

1. 洋蔥、黑橄欖 2 顆、紅椒及青椒均切小切；鮪魚瀝乾油漬；美生菜洗淨濾乾；小黃瓜切片。
2. 將材料 A 拌勻為鮪魚沙拉，美乃滋勿加太多，容易生水。
3. 在法國麵包 2/3 處橫切開，塗上美乃滋，加入生菜、小黃瓜、起司片及鮪魚沙拉，再以黑橄欖及芥末醬裝飾。

1. Chop onion, olives, red bell pepper and green pepper; strain tuna fish; clean and strain lettuce; cut cucumber to be slices.
2. Mix all ingredient A to tuna salad.
3. Cut baguett then spread mayonnaise, add lettuce, cucumbe, cheese slices and add tuna salad, then garnish with black olive and dijon mustard.

總匯三明治
Club Sandwich

難易度 ★　　時間 20 分

材料 Ingredients

6 片	吐司片	pieces of toast slice
4 片	培根	bacon slice
4 片	紅蕃茄片	tomato, slice
2 片	起司片	cheese slice
40g	豌豆苗	peas sprout
20g	美生菜	lettuce
2 顆	蛋	eggs
1 片	水煮雞胸肉	piece of cooked chicken breast
200g	冷凍薯條	frozen French fries
30g	美乃滋	mayonnaise
1 條	黃瓜片	cucumber, slice

作法 Method

1. 將培根煎上色；蛋打散入鍋中煎成蛋皮備用。
2. 用 190℃ 油溫炸薯條至呈金黃色，以適量鹽、胡椒調味。
3. 吐司烤上色塗上美乃滋，分別夾上培根、雞胸肉、蛋為一層；蕃茄、起司片、美生菜、小黃瓜、豌豆苗做為另一層；對角切割成三角形，最後再搭配薯條即可。

1. Pan-fry bacon to be light brown colour; beating egg in a sauce pan and pan-fry to be egg sheet.
2. Deep-fry French fries at 190℃ until golden brown and season salt, pepper to taste.
3. Brown the toast and spread the mayonnaise then put bacon, chicken breast and egg to be the first layer; put tomato, cheese slices, lettuce, cucumber and peas sprout to be another layer; then cut the sandwich diagonal and serve with French fries.

材料 Ingredients			
	3 顆	水煮蛋	boiled eggs
	2 片	全麥吐司	wheal toast
	2 片	白吐司	white toast
A	30g	美乃滋	mayonnaise
	5 片	薄荷葉，切碎	mint leaf, chopped
	2 片	九層塔切碎	basil, chopped

作法 Method

1. 將蛋切碎，加入材料A拌勻。
2. 將蛋沙拉夾入吐司內，再以刀切去吐司邊後切成長方形即可。

1. Chop eggs then mix well with ingredients A.
2. Put the egg salad in toasts then cut off the crusts and cut into 3 or 4 strips.

※ 三明治製作注意事項：
Instruction of making sandwich：

1. 選擇的麵包可有所變化，但麵包要新鮮鬆軟。
2. 生菜應洗淨且瀝乾。
3. 主材料、配料的大小、刀工須一致。
4. 選擇新鮮、好品質的食材。

1. Choose different and fresh breads.
2. Clean lettuce and strain them.
3. Cut same size of ingredients of sandwich filling.
4. Use fresh, high-quality ingredients.

三明治套餐

蛋沙拉三明治
Egg Salad Sandwich

難易度 ★   時間 15 分

夏日水果沙拉三明治
Summer Fruit Salad Sandwich

難易度 ★　　　　時間 20 分

材料 Ingredients

A	2 根	香蕉	bananas
	125cc	優格	yogurt
	20cc	檸檬汁	lemon juice
B	1 顆	火龍果丁	diced dragon fruit
	8 顆	草莓丁	diced strawberries
	2 顆	奇異果丁	diced kiwifruit
	1/4 顆	蜜世界丁	diced melon
C	1/2 顆	蘿蔓生菜	romaine lettuce
	切片	全麥麵包	sliced wheat bread

作法 Method

1. 香蕉優格醬：將材料 A 放入果汁機打成泥狀。
2. 將香蕉優格醬拌入綜合水果材料 B 中攪拌均勻。
3. 將蘿蔓生菜及水果沙拉夾入 2 片全麥麵包中即可。

1. Bananas yogurt: put ingredients A into blender and puree.
2. Mix banana yogurt with assorted fruit from ingredients B.
3. Place romaine and fruit salad into 2 pieces of wheat breads.

Menu For Family Party 家 庭 宴會餐

Mango Fruit Cooler
芒果水果汁

Mixed Mushroom Salad
綜合蘑菇沙拉

Beef Goulash Soup
匈牙利牛肉湯

Gratin Seafood Rice
焗海鮮飯

Spaghetti Bolognaise
義大利肉醬麵

Home Made Coffee Mousse
咖啡慕斯

這是一套屬於全家聚會時開懷暢飲的美味套餐，在用餐時，不僅可增添家庭團聚熱鬧的氣氛與家庭成員間彼此的關懷，更可凝聚每個人的向心力。

這套套餐主要是以自助餐 Buffet 的型態，想像將自助餐移到自家的餐桌上，有好吃的蘑菇沙拉、味道濃郁鮮美的匈牙利牛肉湯、香味四溢的焗海鮮飯、義大利肉醬麵及超優的飯後甜點咖啡慕斯，都是一大碗或一大盤的份量，可隨自己的喜好盛取，輕鬆自在無負擔，再搭配酸酸甜甜的芒果水果汁，好不快活！

This is a set menu for the whole family to get together. Having this dishes will not only increase the happiness of family time but also condense centripetal force of everyone.

This menu is buffet style, Imagine that it seems like you move the buffet into the table of your own house. There are mushroom salad, Beef Goulash Soup, Gratin seafood rice, Spaghetti Bolognaise, coffee mousse for dessert and mango juice. Those are buffet style that you can taste no matter which one you like!

芒果水果汁
Mango Fruit Cooler

難易度 ★　　　時間 15 分

材料 Ingredients

	4 片	柳橙片	pieces of orange slice
A	250cc	紅莓汁	cranberries juice
	20 顆	冰塊	ice rocks
	750cc	柳橙原汁	orange juice
	2 顆	芒果丁（預留 1 杯裝飾用）	diced mangos (keep 1 cup for garnish)
	45g	蜂蜜	honey

作法 Method

1. 將材料 A 用果汁機打勻，倒入果汁杯中。
2. 加上芒果丁，用薄荷葉、檸檬皮裝飾即可。

1. Put ingredients A in blender and blend, and then pout into a glasses.
2. Add mango dice, and then garnish with mint leaf and lemon zest.

綜合蘑菇沙拉
Mixed Mushroom Salad

難易度 ★   時間 30 分

家庭宴會餐

材料 Ingredients

30g	橄欖油	olive oil	
30cc	白酒	white wine	
200g	蘑菇	mushrooms	
150g	生香菇	chinese mushrooms	
150g	鮑魚菇	oyster mushrooms	
2 顆	蒜頭	garlic	
1 支	紅辣椒	red chili	
1 支	綠辣椒	green chili	
20g	九層塔	basil	

醬汁 sauce

60cc	橄欖油	olive oil
30cc	白酒醋	white wine vinegar
20cc	蜂蜜	honey
20g	芥末醬	mustard
適量	鹽、黑胡椒	T.T. salt, pepper

作法 Method

1. 將蘑菇，生香菇及鮑魚菇去梗，切塊；紅綠辣椒切段；蒜頭拍裂。
2. 用橄欖油將三種菇類炒香，再加入辣椒及蒜頭拌炒 1 分鐘，加入白酒縮乾，移火，待冷備用。
3. 將醬汁的材料攪拌均勻成濃稠狀。
4. 再將醬汁拌入蘑菇中，撒上九層塔葉即可。

---

1. Take off the stalk of mushrooms and cut to be big cubes; cut red and green chilies to be strips; crush garlic.
2. Saut?these mushrooms with olive oil then add chili add garlic to fry for 1 minutes, add white wine reduce and turn of the heat, then keep it.
3. Mix up equally the ingredients of sauce to be slightly thickened.
4. Then mix sauce in assort mushrooms and garnish basil leaves.

匈牙利牛肉湯
Beef Goulash Soup

難易度 ★★　　時間 50 分

材料 Ingredients

1/2 顆	洋蔥	onion
2 顆	紅蕃茄，去皮去籽	tomatoes, peeled and seeded
1/2 顆	青椒	green pepper
30g	奶油	butter
5g	檸檬皮碎	lemon zest
5g	蒜頭末	chopped garlic
3g	辣椒粉	chili powder
250g	牛里肌肉	beef loin
1 顆	洋芋	potato
1.5 公升	牛高湯	liter beef stock
15g	匈牙利紅椒粉	paprika powder
30g	蕃茄糊	tomato paste
3g	小茴香粉	cumin
1 片	月桂葉	bay leaf
少許	鹽、胡椒	T.T. salt, pepper

作法 Method

1. 蕃茄、洋蔥、洋芋、青椒、牛肉分別切成小丁；青椒燙煮後沖冷水備用。
2. 將洋蔥炒至呈淡黃色，續入蒜末、檸檬皮末及匈牙利紅椒粉及小茴香粉，再加入牛肉炒至上色。續放入蕃茄糊炒至暗紅色後，加入月桂葉及辣椒粉。
3. 加入牛高湯，以大火煮至沸騰後，續入洋芋丁及蕃茄丁，加蓋以小火燜煮 30 分鐘至牛肉及洋芋熟透，最後加入鹽、胡椒調味，以青椒丁裝飾即可。

1. Cut tomato, onion, potato, green pepper and beef to be dice; blanch the green pepper.
2. Saut?onion with butter until light brown, then add garlic, lemon zest, paprika, cumin and beef keep frying for a minutes and add tomato paste until the golden brown colour then add chili powder and bay leaf.
3. Add beef stock, diced potatoes and diced tomatoes, bring to simmer for about 30 minutes until beef and potatoes are terder, season with salt and pepper, garnish with green pepper.

焗海鮮飯
Gratin Seafood Rice

難易度 ★★　　時間 40 分

材料 Ingredients

	A		
	150g	紅鯛魚片，切丁	red snapper fillet, dice
	150g	花枝肉，切丁	squid, dice
	150g	草蝦仁	shrimps
	150g	生干貝	scallops
	150g	起司絲	grated cheese
	400g	白飯	steamed rice
	125cc	無糖鮮奶油	cream
	65cc	白酒	white wine
	50g	新鮮洋菇（對切）	mushrooms (cut to be two equal parts)
	60g	奶油	butter
	60g	麵粉	flour
	250cc	牛奶	milk
	1/2 顆	洋蔥，切碎	onion, chop

作法 Method

1. 烤箱預溫200℃。
2. 將紅鯛魚、花枝、蝦仁、生干貝及洋菇依序燙煮、過濾，將其濃縮成高湯。
3. 奶油炒香洋蔥碎，續入麵粉拌成麵糊，再入高湯攪拌至無顆粒狀後，依序加入牛奶及白酒拌攪均勻，最後放入海鮮料及鮮奶油煮至沸騰，再用鹽、胡椒調味。
4. 將焗鍋加入白飯墊底，再加入奶油海鮮料，撒上起司絲，送入烤箱以200℃焗烤約15分鐘至呈金黃色即可。

---

1. Preheat oven to 200℃.
2. Blanch with snapper, squid, shrimps, scallops and mushrooms then strain, and reduce stock.
3. Saut?onion with butter, then add flour and make roux, then add stock and until no granule, then add milk and white wine until smooth, add seafood and cream, keep cooking until it is boiled then season with salt and pepper.
4. Put steamed rice on the bottom add seafood ingredients, spread cheese and bake 200℃ for about 15 minutes until done.

義大利肉醬麵
Spaghetti Bolognaise

難易度 ★　　　　時間 60 分

材料 Ingredients

	500g	牛後腿絞肉	ground round beef mince
	500g	去皮蕃茄	peeled tomatoes
	400g	義大利麵條	spaghetti
	100g	蕃茄糊	tomato paste
	75cc	橄欖油	olive oil
	800cc	雞高湯	chicken stock
A	1 1/2 顆	蒜頭	garlic
	7g	匈牙利紅椒粉	Hungarian paprika
	1 顆	洋蔥	onion
	1/2 顆	青椒	green pepper
	2 支	西芹	stalks celery

香料 Spices

1/2 片	月桂葉	bay leaf 2.5g
2.5g	九層塔	basil
2.5g	俄立崗	oregano
1/2 支	紅辣椒	red pepper
25g	起司粉	cheese powder

作法 Method

1. 蒜頭、洋蔥、紅辣椒及九層塔切碎；青椒、西芹、去皮蕃茄切丁。
2. 熱鍋熱油後將牛絞肉炒香並加入匈牙利紅椒粉，炒至金黃色備用。
3. 鍋熱炒香材料 A（勿上色），續入牛絞肉拌炒 3 分鐘，加蕃茄糊，再拌炒至紅褐色，入去皮蕃茄及雞高湯煮滾，最後入香料，以小火慢煮 45 分鐘關火。
4. 義大利麵條以滾水煮約 8~10 分鐘，過濾，拌油，再加入肉醬，灑上起司粉即可。

---

1. Chop up garlic, onion, red pepper and basil. Dice green pepper, celery, and peeled tomatoes.
2. Heat oil and saute ground beef until fragrant, add the Hungarian paprika, and fry to golden brown and set aside.
3. Heat the pot and fry ingredients A (don't brown them), put in ground beef and stir-fry for 3 minutes. Add tomato paste, then stir to reddish brown. Add the peeled tomatoes and chicken stock, and bring to boil. Put in the spices and cook slowly on slow fire for 45 minutes.
4. Blanch spaghetti in boiling water for about 8-10 minutes, then filter and mix with oil. Add the meat sauce and sprinkle with cheese powder.

咖啡慕斯
Home Made Coffee Mousse

難易度 ★★　　　時間 90 分

材料 Ingredients

	45g	即溶咖啡粉	instant coffee
	2 片	吉利丁	gelatin
	30g	咖啡酒	kahlua
	170cc	鮮奶油	fresh cream whipping
A	2 顆	蛋黃	egg yolks
	15g	砂糖	sugar
	30g	水	water
B	2 顆	蛋白	egg white
	30g	砂糖	sugar
	1~2g	鹽	salt

作法 Method

1. 將材料A中的蛋黃、砂糖及水倒入攪拌盆中，充分混合後隔水加熱，將蛋黃迅速攪拌成乳狀。
2. 加入即溶咖啡粉混合。
3. 吉利丁以冷水泡軟瀝乾後，並溶入作法2的材料中，並讓它自然冷卻，加入咖啡酒攪拌均勻。
4. 將材料B的蛋白打至微發後，分2~3次將砂糖加進去，繼續打至全發。
5. 輕輕的混合蛋白及鮮奶油，倒入咖啡慕斯內，注意不要把氣泡弄塌了。
6. 將混合好的慕斯裝進容器內，整平表面，放進冰箱冷藏1小時至其凝固。
7. 再擠上奶油花，用巧克力咖啡豆作裝飾即可。

1. Add egg yolk (ingredients A), sugar and water in a pot then mix well, over hot water of the pot until thicken creamy.
2. Add instant coffee and mix up.
3. Soaked gelatin in cold water to be soft then strain it, put it into the ingredients (step 2) and melded, then add kahlua and mix up equally.
4. Whip egg whites to a soft peak. Add 30g sugar and whip to stiff peak.
5. Fold up egg white and whip and coffee mixture, do not collapse the form.
6. Put the coffee mousse in the container, flat the surface then keep it in fridge for 1 hour until done.
7. Garnish with whipping cream and chocolate coffee beans.

Vegetarian Set Menu 全素套餐

Broccoli and Cauliflower Salad
翠玉花菜沙拉

Cream of Green Pea Soup
奶油青豆仁湯

Vegetable Lasagna
蔬菜千層麵

Strawberry Shortcake
英式草莓鬆餅

一般而言，大多數人吃素是為了宗教信仰，也有人是為了身體健康。對西方人來說，所謂素食，是指不吃肉類。而此套套餐是針對國人的需求而設計，完全不添加蔥、薑、蒜等葷辛調味料的全素菜單。蔬菜中含有多種維生素及礦物質，是大家不可或缺的健康食品。希望這套套餐能提高大家對西式素食的烹調興趣，且帶給大家更均衡的飲食，更健康的生活。

In general, most vegetarism eat vegetarian food for religion or health. The term "vegetarian", for western people, means not to eat meat. This set menu is designed for Chinese which contains no spring onion, ginger, garlic, and so on. Vegetables contain a variety of vitamins and mineral substance; which are essential ingredients to everyone. Hopefully this set menu is able to help people increase the cooking interests of western vegetarian; bring the balanced diet and live a healthier life for people.

翠玉花菜沙拉
Broccoli and Cauliflower Salad

難易度 ★　　　　時間 30 分

材料 Ingredients

250g	青花菜	broccoli
250g	白花菜	cauliflower
80g	杏仁片	almond slices
2 顆	柳橙（去皮，切圓片，預留 1/2 顆壓汁）	
	oranges (peeled, cut to be round slices, keep half orange and press to juice)	
1/2 顆	檸檬（壓汁，預留 3g 檸檬片，切碎備用）	
	lemon (press to be 2 tsp of juice, prepare zest)	
200cc	橄欖油 olive oil	
適量	鹽、胡椒、糖、黑胡椒碎　T.T. salt, pepper, sugar, chopped black pepper	

作法 Method

1. 將青花菜及白花菜切成小朵，燙煮 1 分鐘後過濾沖冷水，備用。
2. 杏仁片放入 180℃烤箱烤約 3～5 分鐘至呈金黃色，備用。
3. 將檸檬皮、檸檬汁、柳橙汁及橄欖油放入果汁機打散成濃稠醬汁，以鹽、胡椒調味，若太酸可加入少許的糖。
4. 再將醬汁淋入花菜中，拌均勻，上盤後再用杏仁片、柳橙片及黑胡椒碎裝飾即可。

1. Cut broccoli and cauliflower to be small pices then blanch.
2. Toast almond slices until the golden brown colour.
3. Put lemon zest, lemon juice, orange juice and olive oil into blender and blend to be dressing, season with salt and pepper.
4. Toss the cauliflower and broccoli with dressing, garnish with almond slices, orange slices and black pepper.

奶油青豆仁湯
Cream Of Green Pea Soup

難易度 ★　　時間 45 分

全素套餐

材料 Ingredients

500g	青豆仁（預留 50g 裝飾用）	green peas (keep 50g for garnish)
2 支	西芹丁	stalk of celeries (diced)
250g	素火腿丁	diced vegetarian ham
1 公升	香菇高湯	litre mushroom stock
125cc	無糖鮮奶油	fresh cream
15g	香菜	coriander
30cc	橄欖油	olive oil
適量	鹽、胡椒	TT. Salt and pepper

香料包 Spices bag

1 片	月桂葉	bay leaf
5 顆	白胡椒粒	white pepper corns

作法 Method

1. 用橄欖油將西芹及青豆炒香。
2. 加入香菇高湯，煮滾後加入香料包再用小火燜煮約半小時，至青豆仁軟化，取出香料包，待冷卻。
3. 用果汁機將青豆仁打成濃稠無顆粒狀。
4. 加入鮮奶油及素火腿丁煮滾，用鹽、胡椒調味。
5. 最後加入香菜及 1/2 杯青豆仁裝飾即可。

1. Sauté with celeries and green peas with olive oil.
2. Add mushroom stock, spice bag bring to boiled then simmer for 30 minutes, take out the spicy bag.
3. Blend soup in blender to be thicken and has no granule.
4. Add fresh cream and vagetarian ham bring to boiled then season with salt and pepper.
5. Garnish with green peas.

蔬菜千層麵
Vegetable Lasagna

難易度 ★★　　　　時間 75 分

蕃茄醬汁 材料　Ingredients of tomato sauce

1500g	去皮蕃茄罐，切丁	peeled canned tomato, dice
125g	紅蘿蔔碎	chopped carrot
1 支	西芹碎	chopped celery
50cc	橄欖油	olive oil
3 片	九層塔碎	pieces chopped basil, choped
500cc	香菇高湯	mushroom stock
少許	鹽、胡椒	T.T. salt, pepper
1 片	月桂葉	bay leaf

蕃茄醬汁 作法　Method of tomato sauce

1. 用橄欖油將紅蘿蔔及西芹炒香後，加入蕃茄丁及香菇高湯煮約 20 分鐘，加入九層塔碎、1 片月桂葉再煮 10 分鐘，最後用鹽、胡椒調味即成蕃茄醬汁。

1. Sauté carrot and celery with olive oil, add diced tomatoes and mushroom stock to simmer for 20 minutes, add basil and bay leaf and keep simmer for another 10 minutes, and season with salt and pepper to be tomato sauce.

材料　Ingredients

12 片	千層麵皮	12 pc lasagna
1500cc	蕃茄醬汁	tomato sauce
120g	蘑菇片	sliced mushroom
1 條	茄子，切丁	eggplant , dice (1 cm)
1 顆	紅甜椒，切丁	red bell pepper , dice (1 cm)
1 顆	黃甜椒，切丁	yellow bell pepper , dice (1 cm)
250g	南瓜，切丁	pumpkin , dice (1 cm)
120g	莫札瑞拉起司片	Mozzarella cheese
250g	起司絲	grate cheese
30cc	橄欖油	olive oil
25g	起司粉	parmesan cheese powder
5 片	九層塔絲	pieces of basil, julienne

作法　Method

1. 烤箱預溫 180℃。
2. 麵皮用熱水燙約 8 分熟，過濾，待冷，拌油。
3. 用橄欖油依序將南瓜，蘑菇、茄子、黃甜椒及紅甜椒炒香，再加入蕃茄醬汁及九層塔，以慢火煮 20 分鐘至南瓜熟透，加鹽、胡椒調味。
4. 將焗烤盤塗上一層油，依序鋪上麵皮、蔬菜醬、起司絲（片）重覆共 3~4 次至 8 分滿。
5. 送入烤箱烤約 20 分鐘至呈金黃色即可。

1. Preheat oven to 180℃ .
2. Blanch the lasagna with for about 8 minutes then strain it, keep cooled and then mix with oil.
3. Saut?pumpkin, mushroom, eggplant, yellow bell pepper and red bell pepper with olive oil then add tomato sauce and basil, simmer for 20 minutes until pumpkin is cooked then season with salt and pepper.
4. Spread oil in baking tray, then spread lasagna, sauce, cheese on it for a layer, make 3 ~ 4 layer.
5. Bake about 20 minutes until golden brown colour.

全素套餐

英式草莓鬆餅
Strawberry Shortcake

難易度 ★★　　　時間 30 分

材料 Ingredients

餡料及醬汁 Stuffing and sauce
- 600g 草莓 strawberry
- 40g 糖 sugar
- 20g 糖粉 icing sugar
- 5cc 白蘭地 brandy

香草奶油 Vanilla cream
- 180cc 打發有糖鮮奶油 whipped cream
- 5g 香草精 vanilla extract

鬆餅 Short cake
- 60g 奶油 butter
- 250g 中筋麵粉 flour
- 15g 泡打粉 baking powder
- 2.5g 鹽 salt
- 45g 糖 sugar
- 180cc 無糖鮮奶油 fresh cream

作法 Method

1. 烤箱預熱 220℃。
2. 草莓洗淨去蒂，取 200g 草莓，用果汁機打成泥狀，再慢慢加入糖粉，攪拌至糖粉溶化，即成醬汁，亦可加入少許白蘭地。
3. 另 400g 草莓對切，用 40g 的糖醃漬 10 分鐘備用。
4. 中筋麵粉加入泡打粉，加入糖及鹽，拌均勻後再加入切小塊的奶油，用手抓奶油成小顆粒，再加入無糖鮮奶油，混合均勻揉成麵糰，用桿麵棍桿成 1 公分厚。
5. 用 3 吋的圓型模，蓋成圓形鬆餅 8～10 個，送入烤箱烤 12～15 分鐘至金黃色為奶油鬆餅，取出備用。
6. 將打發的鮮奶油拌入糖及香草精，拌均勻。
7. 將烤過的鬆餅夾入糖漬草莓及香草鮮奶油，再淋上草莓醬汁即可。

---

1. Preheat oven to 220℃.
2. Clean strawberry, put 200g strawberry in a blender and pure? then add icing sugar slowly and stir it until melted add few brandy to be sauce.
3. Cut the other 400g strawberries then marinate with 40g sugar for 10 minutes.
4. Mix baking powder, flour, sugar and salt and then add diced butter, grab the butter to be small granules by hands, then add 180cc cream, mix up to be sugar dough. Roll out to 1cm thickness.
5. Make 8～10 round sugar dough by 3 inches round model, then bake for 12～15 minutes until done to be shortcake.
6. Mix up whipped cream with vanilla extract.
7. Place the strawberries and vanilla cream in shortcake and close another piece. Serve with strawberry sauce.

BBQ Menu 烤肉餐

Herb, Garlic Scallop and Prawn Skewer
蒜味香料干貝草蝦串

Mexican Steak with Avocado Salsa
墨西哥烤牛排附酪梨醬

Grill Vegetable Kebabs
烤蔬菜串

Corn on the Cob
碳烤原味玉米

Pina Colada Pineapple
椰香烤鳳梨

談到 B.B.Q.，不禁讓人想起秋高氣爽的烤肉季節，其實烤肉餐也可以是有質感的宴客套餐，不論是在室內或戶外，只要有個碳烤台，或是家用烤箱，都能讓您一年四季烤出香味四溢的美食。此套餐前菜為蒜味香料干貝草蝦串，或可依個人喜好更換為淡菜、鮭魚及花枝等海鮮，食材改變，美味不變；主菜則為墨西哥式的碳烤牛排，加上酸辣濃郁的酪梨醬，再配上烤蔬菜串，為健康均衡一下，還沒飽嗎？再來一串碳烤原味玉米，絕對讓您回味無窮，甜點則是具蘭姆酒香的椰香烤鳳梨。啤酒或蘇打飲料更是不可缺少的飲料。

When we mention about B.B.Q. , it make us remind of the B.B.Q. season, actually BBQ meal could be a wonderful entertainment banquet for guests, and every season, you can always make perfect taste if you have a grill stance or oven. The appetizer of this set menu is herbal scallop and shrimp skewer with garlic. The main course is Mexican grilled steak with sour and spicy avocado sauce, then set up with vegetable skewers, it helps to keep healthy. Don't you feel enough? Taking another grilled corn would make you satisfied. Dessert is baked pineapple with coconut milk with Rum. Beers or soda are best drinks.

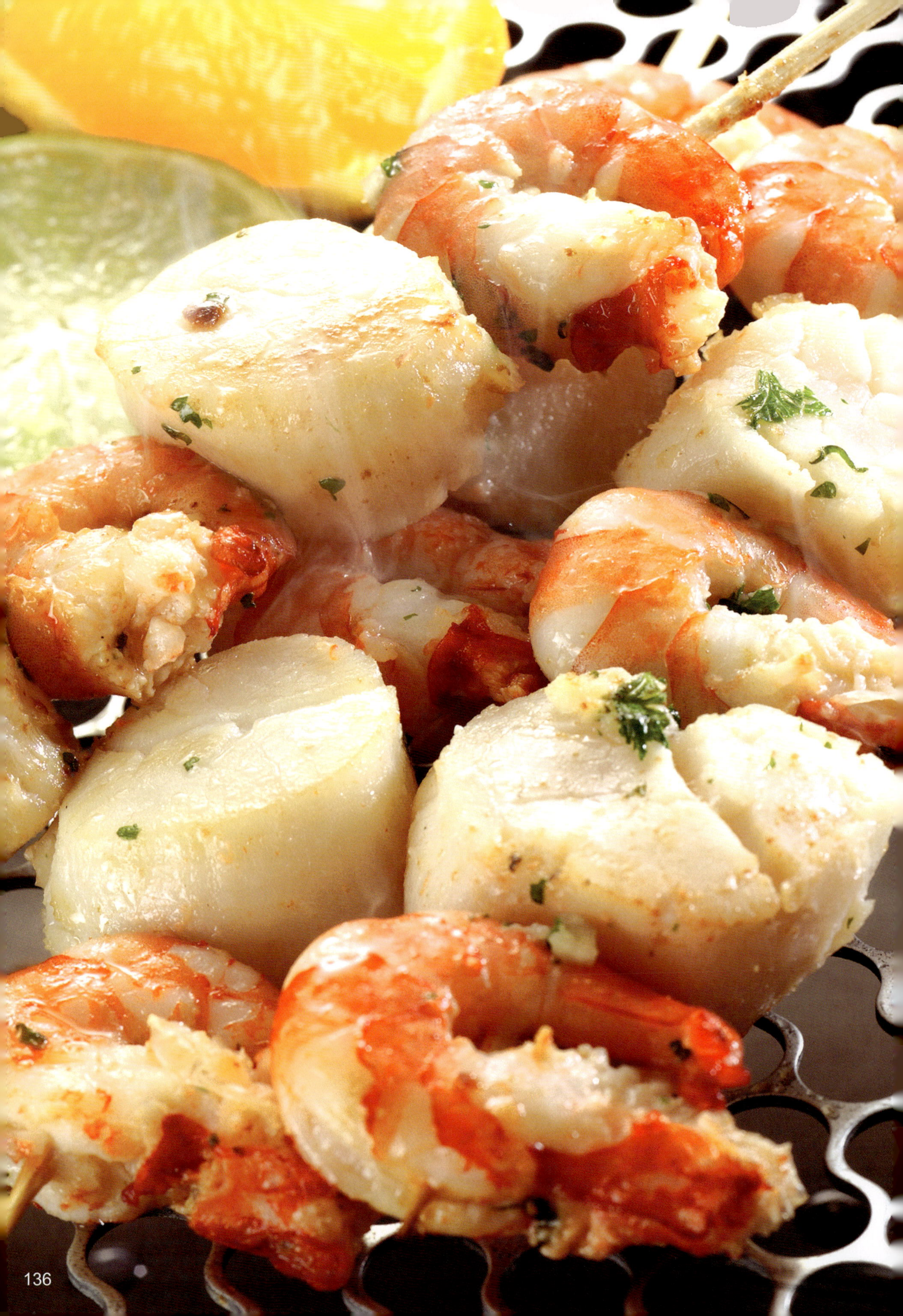

蒜味香料干貝草蝦串
Herb Garlic Scallop and Prawn Skewer

難易度 ★　　　　　　　　　　　　　　　　　時間 50 分

材料 Ingredients

24 隻	草蝦（1 隻約 20g，去頭及殼）	grass shrimps (about 20g each, peeled)
16 顆	生干貝	scallops
20g	巴西利碎	chopped parsley
60cc	檸檬汁	lemon juice
30cc	橄欖油	olive oil
60g	奶油	butter
2 顆	蒜頭碎	chopped garlic
適量	鹽、胡椒	T.T. salt, pepper

作法 Method

1. 將草蝦及干貝加入巴西利、檸檬汁及鹽，以胡椒醃漬約 30 分鐘，備用。
2. 小火加熱橄欖油，放入蒜頭拌炒 30 秒，移火，再加入奶油塊，攪拌至奶油融化。
3. 將醃漬的香料海鮮拌入一半的蒜頭奶油，拌攪均勻。再用竹籤依序將 3 隻草蝦及 2 顆干貝串成 1 串共 8 串。
4. 放在碳火上烤 8～10 分鐘，刷上醃料奶油烤到兩面上色至熟即可。
5. 上盤時，可附另一半的香蒜奶油為沾料。

1. Marinate the grass shrimp, scallops with the parsley, lemon juice, salt and pepper for 30 minutes.
2. Sauté the garlic with olive oil until light brown after mix soft butter to be garlic butter.
3. Marinated herbal seafood and garlic butter, then make 3 shrimps and 2 scallops to be 1 skewer, make 8 skewers.
4. Grill seafood for 8～10 minutes, then brush garlic butter until done.
5. Serve with garlic butter dip sauce.

烤肉餐

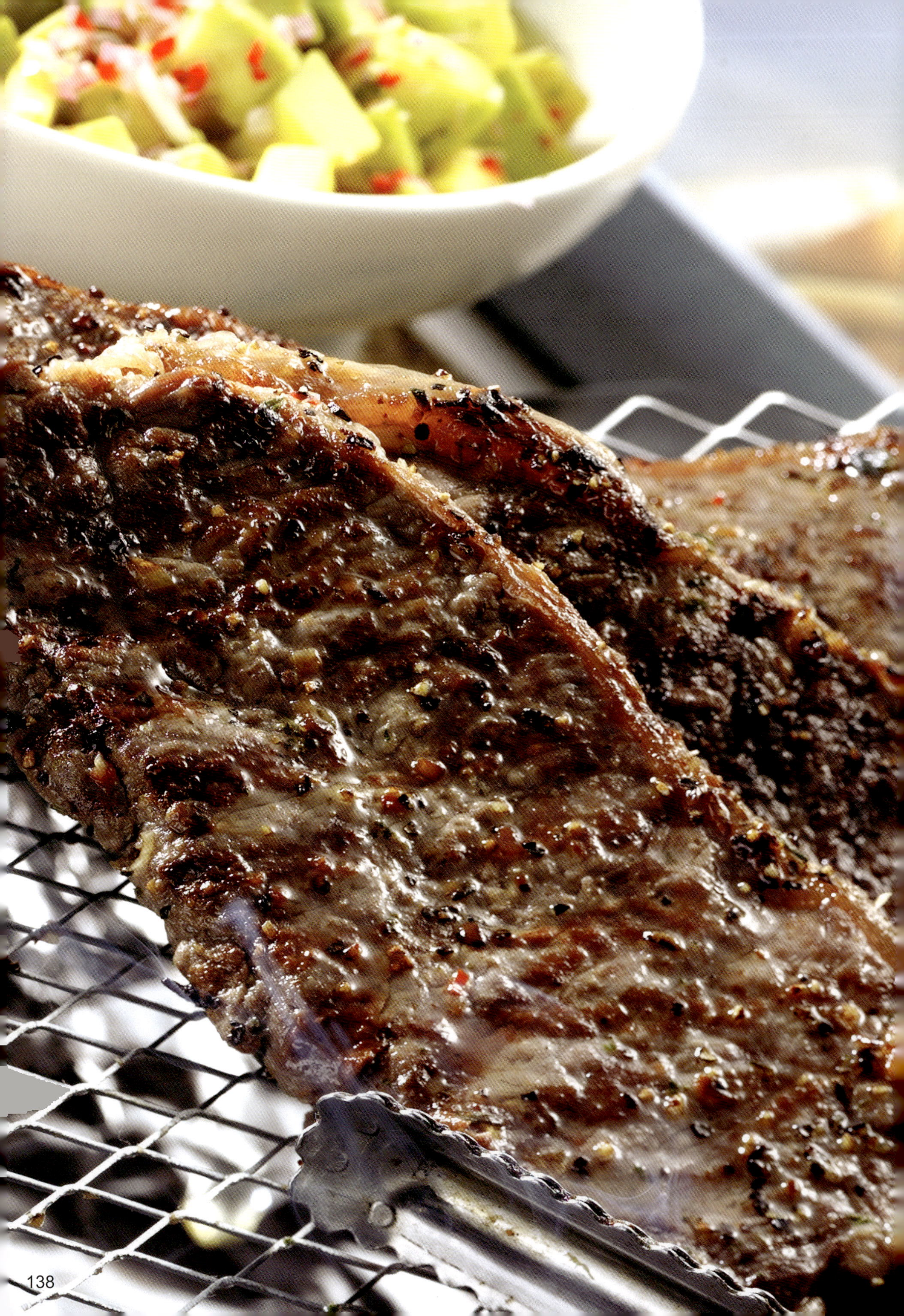

墨西哥烤牛排附酪梨醬
Mexican Steak with Avocado Salsa

難易度 ★★　　　　　　　　　　　　時間 75 分

材料 Ingredients

4 片 120g	沙朗牛排	4 pieces of sirloin steak, 120g each

醃料 Marinate

45cc	蔬菜油	vegetable oil
1/2 顆	紅蔥頭碎	chopped shallot
1 支	紅辣椒（去籽切碎）	chili (seeded and chopped)
1 顆	蒜頭碎	chopped garlic
5g	香菜碎	chopped coriander
2.5g	奧勒崗	oregano
3g	小茴香粉	cumin
適量	鹽、胡椒	T.T. salt, pepper

作法 Method

1. 將牛排拍平，加入醃料拌均勻，醃漬 1 小時。
2. 碳烤牛排兩面約 5 分鐘至 7 分熟。
3. 附上酪梨醬。

---

1. Marinate steak with marinate for 1 hour.
2. Grill of steak to be about mid-well.
3. Serve with avocado sauce.

酪梨醬 材料 Ingredients of avocado sauce

1 顆	酪梨（去皮，切 1 公分丁）	avocado (peeled and cut to be 1cm cubes)
1 顆	檸檬（壓汁，2 片檸檬皮，切碎）	lemon (squeeze to get the juice, 2 pieces of lemon zest, chop it)
1/2 顆	紅洋蔥碎	chopped red onion
1 支	紅辣椒碎	chopped red chili
5g	香菜碎	chopped coriander
適量	鹽、胡椒	T.T. salt, pepper

酪梨醬 作法 Method of avocado sauce

1. 將酪梨丁拌入醃料，攪拌均勻放入冰箱冷藏備用。

---

1. Mix up all ingredients B and season then keep it in fridge.

材料 Ingredients			
1/2 顆	紅甜椒	red bell pepper
1/2 顆	黃甜椒	yellow bell pepper
1/2 顆	青椒	green pepper
1/4 顆	洋蔥	onion
4 顆	小蕃茄	cherry tomatoes
4 朵	生香菇	chinese mushrooms
90cc	橄欖油	olive oil
1 顆	蒜頭碎	chopped garlic
2.5g	百里香	thyme
適量	鹽、胡椒	T.T. salt, pepper

烤蔬菜串
Grill Vegetable Kebabs

難易度 ★   時間 30 分

作法 Method

1. 將紅黃青甜椒切成 2.5 公分大丁；1/2 顆洋蔥開四成半角形；生香菇去梗備用。
2. 橄欖油拌入蒜頭及香料，再用鹽、胡椒調味拌攪均勻，做成香料油。
3. 將所有的蔬菜串成蔬菜串，共 4～6 串。
4. 以中小火碳烤，烤時刷上香料油調味，碳烤 10～15 分鐘至蔬菜兩面上色且熟即可。

1. Cut bell peppers, onion and mushroom to be 2.5cm dice.
2. Mix up olive oil into garlic and herbs, season with salt and pepper to be herbal oil.
3. Skew all vegetables to be vegetable skewers, 4～6 skewers.
4. Grill and brush herbal oil for 10～15 minutes until the both sides of vegetables are golden brown and well-down.

碳烤原味玉米
Corn on the-Cob

難易度 ★　　　時間 45 分

材料 / Ingredients

4 支	帶玉米葉甜玉米	sweet corns with leaves
100g	軟奶油	soft butter
5g	巴西利碎	chopped parsley
1 根	蔥碎	chopped spring onion
5cc	檸檬汁	lemon juice
3g	檸檬皮碎	chopped lemon zest
適量	鹽、胡椒	T.T. salt, pepper

作法 / Method

1. 將玉米葉打開，勿弄斷，去除玉米鬚並洗乾淨，再蓋回玉米葉，在玉米頂部用棉繩綁緊。
2. 以 100℃ 沸水燙煮玉米 5～7 分鐘，過濾備用。
3. 將玉米用中火碳烤約 20 分鐘至熟。
4. 香料奶油：將軟化奶油打散再拌入巴西利碎、蔥、檸檬汁、檸檬皮、鹽、胡椒調味。
5. 當玉米烤熟，再塗上香料奶油即可。

---

1. Open corn, pull out the corn tassel and clean it then cover back the leaf and tie it tight on the top of the corn.
2. Boil corn for 5～7 minutes then strain it.
3. Grill the corn with middle fire for about 20 minutes until it is cooked.
4. Herbal butter: soft butter then mix with chopped parsley, spring onion, lemon juice, lemon zest and season with salt and pepper.
5. Serve with herbal butter.

椰香烤鳳梨
Pina Colada Pineapple

難易度 ★★　　時間 30 分

椰香烤鳳梨 材料　Ingredients of baked pineapple with coconut

1 小顆	鳳梨	small pineapple
50g	奶油	butter
30g	糖	sugar
50g	椰子絲或新鮮椰肉	coconut meat
30cc	蘭姆酒	Rum

椰香烤鳳梨 作法　Method of baked pineapple with coconut

1. 鳳梨洗淨，帶皮及頭一開四，成四塊角形鳳梨，再用小刀取整塊的鳳梨肉，從約 0.5 公分底部劃刀，由頭至尾，每塊角形鳳梨肉再切 6～7 塊。
2. 將奶油融化，將奶油淋在切片的鳳梨上，再撒上白糖。
3. 鳳梨頭部有葉的部分，用鋁箔紙包起，預防碳烤時烤焦。
4. 最後將鳳梨用大火碳烤約 10 分鐘或入 200℃ 烤箱烤 15 分鐘，至鳳梨皮上色，糖溶化。
5. 移火，取出鋁箔紙，淋上蘭姆酒及撒上椰子絲即可。

1. Clean pineapple then cut to be four equal parts, then take out the pineapple flesh, put gentle slits from the bottom (about 0.5cm) to the head, and then cut each part to be 6～7 parts.
2. Melt butter then pour it onto the pineapple wages and then spread sugar.
3. Pack the leaf parts with foil to avoid burned while it is baking.
4. Grill the pineapple with high heat for about 10 minutes or bake 200℃ for about 15 minutes until the skin becomes brown colour and sugar is melted.
5. Take out the foil, then pour rum and spread coconut meat.

Index 索引

沙拉類
鮪魚沙拉附紅椒杏仁醬 16
鳳梨芒果蝦仁 28
高麗菜沙拉 30
凱撒沙拉與香烤雞胸 38
海鮮沙拉盅 48
橙香夏威夷鮭魚沙拉 60
烤鴨香梨華爾道夫沙拉 74
優格馬鈴薯沙拉鱸魚排 88
鮪魚酪梨沙拉 98
綜合蘑菇沙拉 123
翠玉花菜沙拉 130

湯品類
奶油蘆筍湯 18
酥皮奶油玉米湯 27
法式洋蔥湯 40
義式蔬菜湯 49
奶油玉米甜瓜湯 58
蘑菇蓮子卡布奇諾湯 72
麵包甜玉米巧達湯 92
蛤蜊巧達湯 106
奶油白花菜湯 116
匈牙利牛肉湯 124
奶油青豆仁湯 131

主菜類
迷迭香烤雞附原味肉汁 20
夏威夷核果羊小排 22
漢堡排附薯條 30
奶香培根義大利麵 32
檸檬鮭魚捲與奶油芥末醬 44
碳烤菲力牛排與黑胡椒醬 50
藍帶雞排附香料蕃茄 52
香烤果餡火雞捲 62
焗奶油明蝦生蠔 66
豬肋排附美式烤肉醬 68
青醬 75
義式蕃茄海鮮鍋 76
無骨牛小排附蘑菇白蘭地醬 78
水果燕麥片 85
火腿乳酪綜合蛋捲 86
優格馬鈴薯沙拉鱸魚排 88
香煎早餐腸、培根及火腿 95
焗海鮮飯 125
義大利肉醬麵 126

蔬菜千層麵 132
蒜味香料干貝草蝦串 136
墨西哥烤牛排附酪梨醬 138

鹹點類
鄉村洋芋泥 19
餐包 37
黑胡椒牛肉乳酪烘蛋 42
法國吐司 83
焗奶油菠菜烘蛋 94
乳酪雞肉派 100
火腿蒜苗塔 102
乳酪烤淡菜 108
海鮮披薩 109
鮮蝦酪梨三明治 110
香酥雞肉條與炸起司條附塔塔醬 112
鮪魚潛水艇三明治 117
總匯三明治 118
蛋沙拉三明治 119
夏日水果沙拉三明治 120
烤蔬菜串 140
碳烤原味玉米 141

甜點類
皇家巧克力慕斯 24
焦糖布丁 34
英式麵包布丁 46
提拉米蘇 54
肉桂紅酒梨 57
酪梨香草雪球 64
水果巧克力火鍋 70
金桔奶酪 80
美式鬆餅附楓糖醬 91
英式比士吉 101
蘋果馬芬蛋糕 104
咖啡慕斯 128
英式草莓鬆餅 134
椰香烤鳳梨 142

飲料類
德國聖誕紅酒 59
熱巧克力咖啡拿鐵 84
活力蘿蔔汁 84
楓香奶昔 96
香蕉奶昔 114
芒果水果汁 122

總鋪師 3
An Expert
Of Western Cuisine
西餐料理一把罩 精進版

國家圖書館出版品預行編目(CIP)資料

西餐料理一把罩(精進版)/周景堯著. -- 一版.
-- 新北市:三藝出版有限公司, 2021.09 144 面；
19x26 公分. -- (總鋪師；3)
ISBN 978-626-95136-0-4(平裝)

1. 食譜 2. 烹飪

427.12　　　　　　　　　　　　　110015730

作　　　者	周景堯
總 編 輯	薛永年
美術總監	馬慧琪
文字編輯	蔡欣容
攝　　　影	宋棋城、范群浩
出 版 者	三藝出版有限公司
	電話：(02)8521-3848
	傳真：(02)8521-6206
	Email：8521book@gmail.com
	(如有任何疑問請聯絡此信箱洽詢)
	網站：www.8521book.com.tw
印　　　刷	鴻嘉彩藝印刷股份有限公司
業務副總	林啟瑞 0988-558-575
總 經 銷	紅螞蟻圖書有限公司
	臺北市內湖區舊宗路二段 121 巷 19 號
	電話：(02)2795-3656
	傳真：(02)2795-4100
網路書店	www.books.com.tw 博客來網路書店
出版日期	2025 年 8 月
版　　　次	一版二刷
定　　　價	380 元
I S B N	978-626-95136-0-4

上優好書網　　LINE 官方帳號　　FB 粉絲專頁　　YouTube 頻道

Printed in Taiwan
本書版權歸三藝出版有限公司所有 翻印必究
書若有破損缺頁 請寄回本公司更換